The Agricultural Development of Turkey

Oddvar Aresvik

The Agricultural Development of Turkey

PRAEGER SPECIAL STUDIES IN INTERNATIONAL ECONOMICS AND DEVELOPMENT

Praeger Publishers New York Washington London

Library of Congress Cataloging in Publication Data

Aresvik, Oddvar, 1915-
The agricultural development of Turkey.

(Praeger special studies in international economics and development)
Bibliography: p.
1. Agriculture—Economic aspects—Turkey. 2. Agriculture and state—Turkey. I. Title.
HD2037.A7 338.1'09563 74-3575
ISBN 0-275-28851-X

PRAEGER PUBLISHERS
111 Fourth Avenue, New York, N.Y. 10003, U.S.A.

Published in the United States of America in 1975
by Praeger Publishers, Inc.

Printed in the United States of America

To my wife, Gunnvor

PREFACE

This book is based on a study I undertook in 1972 of the introduction of improved wheat technology in Turkey that was financed by the International Maize and Wheat Improvement Center (CIMMYT) in Mexico. The study was based partly on previous visits to Turkey and my writings on the agricultural development of Turkey when I was a Ford Foundation employee from 1963 to 1973, for the last four years working as Agricultural Economist in the regional Arid Lands Agricultural Development Program, with headquarters in Beirut. I was seconded from this program by CIMMYT to work on the wheat technology study in Turkey during the last leg of my contract period.

The analysis of the agricultural development of Turkey was completed during the second half of 1972 and was slightly revised for publication in the beginning of 1974, while I was an employee of the International Bank for Reconstruction and Development working as Agricultural Economics adviser to the Government of Jordan. It aims to present a fairly complete description of the structure and performance of Turkish agriculture in recent years, to locate the main factors in the development process, and to pinpoint the major problems and the main opportunities for the 1970s.

The approach is based upon the hypothesis that, given the natural resources for agricultural development, the land tenure characteristics, the factor prices, and the importance of the risk factor, the rate of growth will depend upon the availability of a stream of well-adapted, high-yielding and profitable technology consistent with environmental and economic conditions, together with the required inputs and credits for financing and successful production campaigns. Since new improved technology is either produced or tested by the Turkish agricultural research service and introduced through the agricultural extension service, the agricultural development services and educational institutions play a central role. They are therefore discussed in detail, and the improvements needed to realize the growth potential are indicated. Since the development strategy has to be tailored to the characteristics of the rural community, these characteristics also have been emphasized.

The tests of the hypothesis are based upon comparisons of the levels of productivity, the quantity of inputs used, and the improved technology introduced in various parts of the country and at different points in time. There are great differences in resource endowments, in the risk factors, and in the availability of well adapted improved technology among the various parts of Turkey. These differences have been used to illustrate the interaction of resource endowments, the importance of risk and uncertainty, the availability of improved technology, and technical change and productivity growth in agriculture.

The introduction of improved wheat technology in the coastal areas of Turkey represents a very good example of how rapid adoption can take place through a strong, well-planned effort. Through its emphasis on this example the study has become somewhat biased toward wheat. However, considering that wheat accounts for about 55 percent of the cultivated land in Turkey, it may be said that the study is not unduly biased. The performance of the new Mexican wheat varieties has greatly stimulated both adaptive research in agricultural production technology and interest in introducing improved technology in Turkey.

The Turkish republic, established in 1923, has been firmly committed to the achievement of modern Western civilization. The development that has taken place is astonishing; and modernization of agriculture has been a very important part of the process. Industrialization alone would provide a far too narrow base and would leave large rural areas behind. This analysis of the agricultural development of Turkey seeks to analyze the ongoing transformation of agriculture, as a reference for institutions, agencies, or individuals that are interested in the economic development of Turkey. Although the book emphasizes the technological and economic aspects of agricultural development, an effort has been made to also cover institutional, sociological, and other aspects.

Because of its successful industrial and agricultural development, and its progress toward self-sustaining economic growth and toward establishing the democratic process, Turkey is a very interesting and inspiring country to study. Agriculturally Turkey is a sleeping giant. Through full use of water resources for irrigation and development, and through the introduction of highly adaptive, improved technology for irrigated as well as rainfed agriculture, production can be increased tremendously.

For the most part the data used in this study come from readily available, published material supplemented by information contained in unpublished reports or obtained through personal interviews.

I am indebted to the Ford Foundation, to the Arid Lands Agricultural Development Program, and to the International Maize and Wheat Improvement Center. Also I am indebted to a number of persons, who provided information and comments that were most helpful. I am especially indebted to Dr. Owen L. Brough, Jr., who collaborated closely in an early draft on Turkish agriculture. Among a great number of other persons I am very much obliged to Dr. Don Winkelmann, who conducted the extensive CIMMYT study on the spread of improved technology for maize and wheat covering several countries, of which the Turkish study to get a better understanding of Turkey's wheat economy was a part, to Mr. Ekrem Gunay, Dr. Ahmet Demirlicakmak, Dr. Resat Aktan, Dr. Nazmi Demir, Dr. Bill C. Wright, Dr. Warren E. Kronstad, Mr. Leonard H. Otto, Dr. Lowell S. Hardin, Mr. James Curley, and Mr. Allan Gall.

I am also grateful to Mr. Sami Rashed, who assisted me in producing the tables, to Mrs. Nada Khattar and Miss Afaf Rashed, who did most of the typing, and to Miss Lana Fawaz Maher Birmamet, who helped me with the last revision.

All of the remaining shortcomings are my own responsibility and I also claim sole responsibility for all matters of interpretation, judgment, and recommendation in this work, which should in no way be taken as representing the views of either the Ford Foundation or the International Maize and Wheat Improvement Center or the World Bank.

CONTENTS

Chapter		Page

LIST OF TABLES

LIST OF FIGURES

GLOSSARY AND EQUIVALENTS

List of Abbreviations

AID	Agency for International Development
CENTO	Central Treaty Organization
CIMMYT	International Maize and Wheat Improvement Center
DSI	General Directorate of State Hydraulic Works
EEC	European Economic Community
EPA	European Productivity Agency
ESSCB	Economic and Social Studies Conference Board (Istanbul)
FAO	Food and Agricultural Organization
GDP	Gross domestic product
GNP	Gross national product
GOT	Government of Turkey
IBRD	International Bank for Reconstruction and Development
IRRI	International Rice Research Institute
OECD	Organization for Economic Cooperation and Development
OEEC	Organization for European Economic Cooperation
SPO	State Planning Organization
TL	Turkish lira = 100 Kurus
TMO	Toprak Mahsulleri Offisi (Soils Products Office)
TSHP	Turkish Society for Housing and Planning

USDA	United States Department of Agriculture
USFOA	United States Foreign Operations Administration

Glossary of Turkish Words

Gecekondu = settlement of houses "raised overnight" without permission

Seriat = the Islamic legal system

Muhtar = village headman

Hoca = imam (prayer leader) of a mosque

Ciftlik = farm

Metric System Equivalents

<u>Linear measures</u>

1 centimeter	= 0.3937 inch
1 meter	= 100 centimeters = 3.281 feet = 1.0936 yards
1 kilometer	= 1000 meters = 1093.6 yards = 0.62137 mile
1 inch	= 2.54 centimeters
1 yard	= 0.914 meter
1 mile	= 1.6093 kilometer

<u>Square measures</u>

1 sq. meter	= 10.764 sq. feet = 1.196 sq. yards
1 sq. kilometer	= 100 hectares = 0.386 sq. miles
1 sq. yard	= 0.836 sq. meters
1 sq. mile	= 2.5899 sq. kilometers

<u>Measures of weight</u>

1 kilogramme	= 1000 grammes = 2.20462 lb.
1 metric ton	= 1000 kilogrammes = 2204.62 lb.
1 lb.	= 0.4536 kilogrammes
1 quintal	= 100 kilogrammes = 220.462 lb.

Measures of land

1 decare (donum)	= 0.2474 acres
1 hectare	= 2.47 acres
1 sq. kilometer	= 0.3861 sq. mile
1 acre	= 0.40468 hectare
1 sq. mile	= 2.59 sq. kilometers

Approximate metric equivalents

1 decimeter	= 4 inches
1 meter	= 1.1 yards
1 kilometer	= 5/8 of a mile
1 decare (donum)	= 1/4 acre
1 hectare	= 2.5 acres
1 hectolitre	= 2-3/4 bush. = 100 litres
1 kilogram	= 2-1/5 lb.
1 metric ton	= 2,200 lb.

Money Equivalents

1.00 Turkish Lira = 100 kurus
13.85 Turkish Lira = 1 U.S. dollar

The Agricultural Development of Turkey

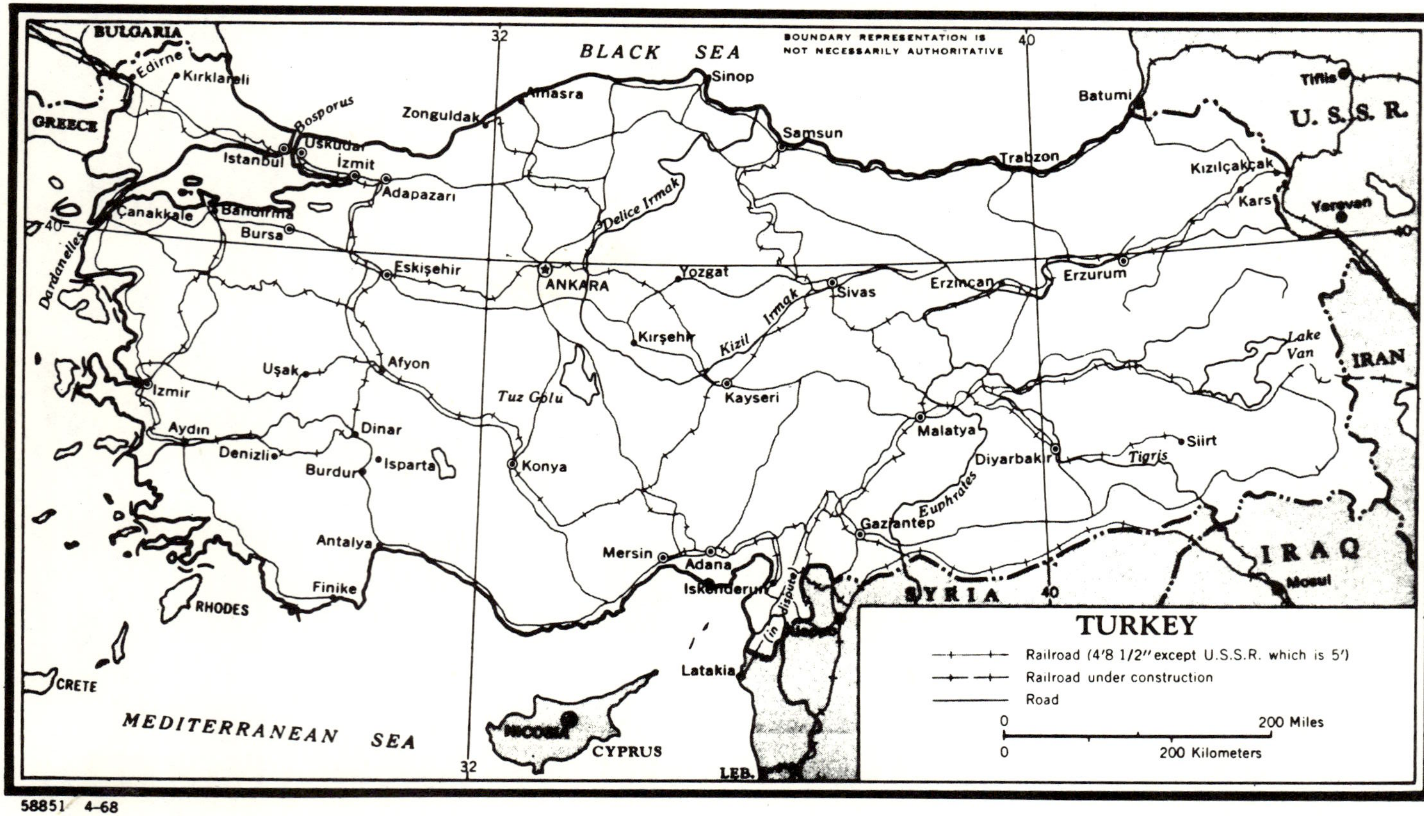

Source: Edwin J. Cohn, Turkish Economic, Social and Politcal Change (New York: Praeger, 1970). Reprinted with permission of the author.

CHAPTER

1

GENERAL INTRODUCTION

TRANSFORMATION OF TURKEY INTO AN INDUSTRIAL STATE

The problems of transforming a traditional agrarian country into a modern industrial state arose in Turkey at least a century later than in the countries of Western Europe. From the beginning of its transformation Turkey was faced with the choice of using state-operated industry or private industry as the main instrument of economic development. It is not easy to categorize the development of the Turkish economy, but three growth periods seem to be generally accepted: (1) the "liberal period," which runs from 1923 until 1929; (2) the "etatist period," from 1930 to World War II; and (3) the "period of interventionism," since World War II. Some writers add the "free enterprise period," beginning with the rise to power of the Democratic Party in 1950, and followed by the "planned development period," beginning in 1963 with the implementation of the First Five-Year Development Plan (1963-67).

State interference in the economy was minimal during the 1920s, though the period was characterized by a number of attempts to legislate cultural and social change. In the 1930s decisive action was taken in favor of extensive state involvement in the economy, in an effort to increase the pace of development. In 1930 and 1931, amid growing disillusionment with private enterprise and partly as a reaction to the worldwide depression, the government and the Republican Peoples Party adopted a program calling for the active intervention of the state in the economy. The purpose of this new policy, known as "etatism," was not the elimination of private enterprise--although it tended to be discouraged--but rather the initiation by the state of projects which were considered vital for national development and defense, but for which private enterprise and capital were considered too weak to be effective.

With its First Five-Year Plan, prepared in 1932 and approved in January 1934, Turkey became the first of the less developed countries to conduct an experiment in planned development. The five-year plans of the 1930s were primarily industrialization plans and gave preference to industry over other economic sectors. The first plan emphasized state entrepreneurship in establishing the major new industries and state control through public financial institutions and agencies. In keeping with its new economic policy, the government established a network of monopolies and state banks, founded a number of basic industries and marketing companies, and nationalized mining and public services. The second plan was prepared in 1936 and approved in 1938, but the war interrupted the implementation of most of its programs.

The major aim of the first plan, which was financed almost entirely from domestic resources, was the development of consumer industries, chiefly textiles but also paper, glass, and ceramics, and of basic industries such as iron, steel, and chemicals. There is little doubt that Turkey, under etatism, did make substantial progress, although the program had serious defects in economic planning. Probably the most serious defect was the almost-complete neglect of agriculture. Development of industry was emphasized at the expense of agricultural development, and little was done to improve the standard of living of the peasants, who constituted about 80 percent of the total population. Illiteracy and general cultural backwardness persisted in the villages. However, the republican-etatist regime consolidated the young state and laid the basic foundations for later social and economic advances.

During World War II Turkey, like most other countries, resorted to far-reaching governmental interference in all fields of economic activity. Instruments of public control extended throughout the economy during the late 1940s. A Third Five-Year Plan, the "Turkish Plan of Economic Development," was prepared in the summer of 1946.

A democratic regime replaced the republican regime in 1950, after free general elections. The democratic government challenged the etatist regime and state-monopoly tendencies with a commitment to gradual transfer of state enterprises to private hands. In practice, however, there was little change in public ownership. The state monopolies retained their position in vital fields, and only limited transfer of public enterprises to private hands took place. At the same time large-scale public investment was devoted to electric power production and highway construction. The pace of development was greatly increased, and more attention was given to agriculture. However, growth outran the available resources, and this led to serious difficulties.

An army revolt in mid-1960 ended the rule of the Democratic party and the military government that was set up soon reintroduced planning as a guiding principle of economic policies. In September 1960 a law was issued that made the State Planning Organization responsible for the preparation of both five-year plans and annual plans within the framework of the five-year plan. This organization is now an advisory body under the authority of the prime minister.

The so-called First Five-Year Development Plan (1963-67) became operative on January 1, 1963, after its approval by the Council of Ministers and the Grand National Assembly. The reactions to the First Five-Year Plan on the whole were quite favorable.

The planning principle was in fact built into the new Turkish constitution. As determined by the High Planning Council, the State Planning Organization operates with 15-year perspective plans, 5-year plans, and annual plans. The annual programs, while designed to implement the annual targets of the five-year plan, also aim at corrections and revisions of the plan in the light of experience gained during the implementation process.

Europe is responsible for about 75 percent of both Turkish imports and exports. An agreement was signed between the European Economic Community (EEC) and Turkey in September 1963, under which Turkey obtained tariff reductions on its main agricultural exports to Common Market countries. On November 23, 1970, an additional protocol was signed between Turkey and the EEC regarding Turkey's entrance and integration into the EEC over a 22-year period.

In 1962 an aid consortium for Turkey was established by most of the Western European countries and the United States. The World Bank joined this consortium in 1965. The USSR, which in the 1930s played a prominent role in Turkey's five-year plan, dropped its assistance later for political reasons, but appeared again in 1972 with both technical and financial aid. External debt services have constituted about 20 percent of Turkey's export earnings. Because of the difficulties of meeting her debt services commitments, "rollback" arrangements were made with the consortium in 1965 for the 1966-70 period, thus reducing the annual commitments considerably by extending the repayment periods.

GROWTH OF TURKISH ECONOMY SINCE 1948

The Turkish economy has had a relatively high growth rate since 1948, as shown in Table 1.1. From 1948 to 1971 the annual growth rate of Turkey's gross national product (GNP) was 5.5 percent and the annual per capita growth rate was 2.9 percent. The agricultural growth rate of 2.9 percent is also a relatively good average for such a long period. The relative importance of agriculture has declined steadily, its share of GNP shrinking from 45 percent in 1948 to 26 percent in 1971. The agricultural population has been growing by 1.7 percent, and consequently the growth rate of value added per capita in agriculture has been only 1.2 percent. The percent of total population employed in agriculture has declined from 75 to 61 percent. There has been a considerable migration of rural families to the cities, which can be partly explained by the unfavorable ratio between nonrural and rural income per capita.

The increase in the value added in agriculture can be partly accounted for by a growth rate for area cultivated of 2.5 percent. A considerable amount of this growth has resulted from bringing more marginal land under cultivation, which could not be expected to result in a proportional increase in production--more so since it has led to a strong decrease and deterioration in the grazing area, which shows a negative growth rate of 3.1 percent.

Inputs from outside the Turkish agricultural sector have played an important role. Both the number of tractors and the consumption of fertilizers have a compound average growth rate of around 20 percent, and for shorter periods the growth rate is much higher, which indicates that the agricultural sector has been quite responsive to these inputs. The relatively high growth rates for the production of such industrial crops as cotton and sugar beets--10 and 7 percent, respectively--and similar growth rates in the area of fruits and vegetables indicate the

TABLE 1.1

Turkish Growth Indicators, 1948-71

	1948	1971	Percent Increase	Annual Compound Rate of Growth (percent)
GNP (billions of TL, 1961 prices)	27.5	94.3	243	5.5
Population (millions)	20.0	35.7	79	2.6
Per capita GNP (TL, 1961 prices)	1,375	2,642	92	2.9
Value added in agriculture (billions of TL, 1961 prices)	12.5	24.1	93	2.9
Agriculture as a percentage of GNP	45.4	26.0	n.r.	n.r.
Population in agriculture (millions)	15.0	21.9	46	1.7
Value added in agriculture per capita (TL, 1961 prices)	83.3	110.0	32	1.2
Percentage of total population in agriculture	75.0	61.0	n.r.	n.r.
Ratio between nonrural and rural income per capita	3.7	4.4	n.r.	n.r.
Area cultivated (million hectares)	13.9	24.4	76	2.5
Area in pastures and meadows (million hectares)	38.0	19.0	n.r.	-3.1
Area sown in cereals (million hectares)	8.1	13.8	70	2.3
Number of tractors (thousands)	1.8	116.0	n.r.	20.0
Consumption of fertilizers (thousand metric tons of nutrients, 1950-54 to 1971)	13.6	494.6	n.r.	21.0
Cereals production (million metric tons)	9.0	19.6	118	3.4
Cotton lint production (thousand metric tons)	58.0	499.0	760	9.8
Sugar beet production (thousand metric tons, 1948-52 to 1971)	966.0	4,254.0	340	7.3
Area in fruits and vegetables (thousand hectares, 1952 to 1971)	636.0	1,500.0	136	4.6

Note: n.r. = Heading not relevant.

Source: Calculations by the author on basis of data from the State Institute of Statistics and the State Planning Organization.

importance of changes toward more intensive, high-value crops. As can be seen from Table A.3 in the statistical appendix, the dominant role of agricultural exports has continued at the same high level of around 80 percent of total Turkish exports.

The development of an all-weather national road network initiated in 1947, and later supplemented by feeder roads, has exerted a profound influence on agricultural development. In 1947 Turkey only had 12,000 kilometers of all-weather roads of low standard. By 1967 a 42,000 kilometer high-standard, all-weather national highway network had been constructed, resulting in an annual growth rate for road transport of around 15 percent. This development expanded the market areas for farm products and specialized production received a great stimulus. Quick, inexpensive, and reliable movement both of inputs to the farming areas and of fresh fruits, vegetables, and livestock products to the markets was made possible. Thus while farmers have become increasingly dependent on inputs produced outside the agricultural sector, such as fertilizers, pesticides, and tractors, they have also become more market oriented. The continually growing level of demand and favorable prices resulting from increased prosperity and a rapidly expanding urban population has stimulated and supported agricultural production, especially in products such as fresh fruits and vegetables and various livestock products. Price support has been an important factor for cereals, tobacco, tea, sugar, hazelnuts and other products.

Part of Turkey's industrial development has supported the agricultural growth. For example, sugar refineries have aided sugar beet production, the development of the textile industry has helped cotton production, the dairy plant industry has boosted milk production and the fertilizer factories and the mechanical workshops have helped to supply additional inputs and services. At the same time the expansion of education and the resulting rise in the literacy rate, together with the introduction of the radio and the improvement in transportation, have broadened the Turkish farmers' horizon and increased their responsiveness to improved technology.

Institutional improvements in the form of more efficient agricultural research, extension, and credit facilities have also played their part; and investments in irrigation systems have been of growing importance, in spite of shortcomings at the farm level.

Land reform laws have been of limited scope and effect and not much land has been distributed under such laws. However, the fragmentation of land has been reduced to a considerable degree.

Agricultural development could most likely have been even more favorable if higher priority had been given to agriculture. There has been a tendency among politicians, civil servants, and intellectuals to give low priority to agriculture and to place industrialization in the forefront, as they have done in so many other less developed countries.

Through expansion of the cultivated area the amount of arable land per person employed in agriculture has remained relatively stable at the level of 1.2 hectares. However, the prospects for expansion of the cultivated area have more or less come to an end. A continuing increase in the agricultural labor force will lead to a declining acreage of arable land per person and most likely to a decrease in the size of the average farm. A further increase in landless labor, which already numbers between one and two million persons in a total agricultural labor force of about 10 million, can also be expected. The data in Table 1.2 give indexes of food production and overall agricultural production on a total basis and a per capita basis during the 1960s.

The table shows that for the past decade agricultural production per capita has been rather static. The same tendency is more pronounced for food production per capita. On the average the food production growth rate in recent years has equaled the population growth rate (2.9 percent), while the growth rate of demand is estimated at about 4.0 percent per annum. Livestock production has shown particularly slow growth, according to the Food and Agricultural Organization (The State of Food and Agriculture, 1970): for the period from 1961 to 1968 the growth rate was only 1.1 percent per annum. Shortages of grazing and forage and excessive feed prices in relation to prices paid for animal products are basic constraints on increased livestock production.

The relatively slow increase in agricultural production and food production in relation to demand has created some strains on the economy. While food grains have been imported under concessional terms, under PL 480, this has not been the case with meat. As a consequence the real price of meat has been rising rather rapidly. The rising real prices of meat cut off the less effective demand and

TABLE 1.2

Indexes of Food Production and Total Agricultural Production in Turkey, 1961-70

Year	Total Agricultural Production	Per Capita Agricultural Production	Total Food Production	Per Capita Food Production	Annual Percent Increase in Value Added in Agriculture (1965 prices)
1961	127	104	128	105	3.2
1962	133	107	134	107	5.8
1963	140	110	140	109	7.6
1964	146	111	140	107	0.0
1965	141	105	137	102	3.3
1966	157	114	151	110	11.4
1967	160	113	152	108	0.9
1968	167	115	160	111	1.9
1969	166	111	161	108	0.1
1970	171	112	166	108	2.0

Source: FAO, The State of Food and Agriculture, 1971.

at the same time impose increasing hardships on the poorer classes of the population, who have to reduce their consumption of valuable animal protein.

Until the early 1960s the country's food production more or less kept pace with the growing demand generated by both the rapidly rising population and moderate increases in per capita income. Increased food production was achieved primarily by increasing the acreage of cultivated land and the number of animals, and to a lesser degree by increasing crop yields and livestock products. This method of increasing food production has now reached the peak of its potential because of limited land resources. Future increases in production must necessarily derive primarily from increased yields and efficiency in production.

The present situation with rising meat prices represents a great challenge, since the current rate of animal protein consumption is relatively low, and represents a decline of one-third over the past two decades as can be seen from Table 1.3.

TABLE 1.3

Estimated Per Capita Protein Content of National Average Food Supply in Turkey (grams per day)

	Total Protein	Animal Protein
Prewar	81.0	15.9
1948-50	80.9	15.3
1964-66	77.9	11.1
1969	76.3	10.7

Source: FAO, The State of Food and Agriculture, 1970.

Wheat is the main source of vegetable protein. This predominance of a cereal grain suggests the need for considerable dietary improvement. Yet the present shortage of meat, along with the rising real prices, will forestall such an improvement.

The tendency toward rising real prices of meat may also have negative repercussions on general economic and social development. Further it will increase the difficulties in adjusting the economy during the transitional period of Turkey's projected entrance into the EEC as a full member.

There are, however, a number of bright spots in the agricultural picture. Yields of wheat and other cereals can be dramatically increased; a reduction in the acreage under fallow and an increase in the production of fodder legumes for sheep and cattle may occur. These factors, togehter with a continued increase in the availability of irrigation and improved water management may reverse the present situation, thus laying the base for diversification and a rapid increase in agricultural production.

IMPROVEMENTS IN WELFARE

Improvements in welfare have been achieved in Turkey concomitantly with the increased production in industry, agriculture, and other sectors. Between 1948 and 1968 the average expectation of a person's life continuing past 5 years climbed from about 53 years to 64 years. Per capita consumption of key commodities also increased rapidly over this period, with consumption of paper more than doubling, of cotton textiles more than tripling, of steel almost quadrupling, and of electricity more than quintupling. Housing conditions have improved among the poorer people, and better clothing has become available to them. The availability of such services and amenities as schools, health facilities, drinking water, and roads has also expanded rapidly in rural areas.

There are still some dark spots in the picture. Income distribution is less equal than it is in most developed countries. In the early 1960s the average annual income for agricultural wage earners was reported to be TL 850, compared with average annual earnings for manual industrial workers of TL 5,210. The tax system appears for the most part either regressive or only slightly progressive and does not tend to reduce disparities between income groups. Agricultural income remains virtually untaxed and similar conditions apply to property owners. Under such conditions large differences occur between the dispensable incomes of large farmers and small farmers.

A recent study indicates that in some areas agricultural development has led to a deterioration in the conditions of the poorer segment of the agricultural population, especially among landless labor: "It appears that living conditions are worst in these villages where production is most developed from a technological point of view."*

There are great differences in levels of development and income among various regions of the country, especially between the western and eastern areas. The Second Five-Year Plan included among its priorities improved food supplies and health services, greater rural development, and the reduction of income differences among social groups as well as among regions. The reduction of income disparities in the agricultural sectors of the country will not be an easy task, since the western coastal regions appear to have the best prospects for future development.

Housing conditions in the rural areas are in need of further improvement. According to the 1965 population census, about 80 percent of the rural houses had no kitchen and 27 percent had no toilet. Only about 3 percent had electricity and running water. In 1964 per capita electric power consumption in Turkey was 142 kilowatt hours, which was about 4 percent of the average for developed countries. About 200,000 dwellings had no windows. There were about 2.7 persons per room in the rural areas compared with 2.52 persons per room for the country as a whole; average number of rooms per rural dwelling was 2.24. And the use of animal dung for fuel was common. Among the improvements, windows, running water, and electricity have been prominent.

Low-income groups that move to the cities must often live in poor housing located in so-called gecekondu areas (settlements of houses "raised overnight" without official permission). Apart from these areas 27 percent of the urban dwellings had no kitchen, 7 percent had no toilet, 49 percent no bath, 26 percent no electricity, and 45 percent no running water, according to the 1965 census.

The deficiencies in housing conditions were often accompanied by disease and undernourishment, with more than 2 percent of the population afflicted with tuberculosis and about 7 percent with trachoma. Bilharziasis seems to be widespread in the irrigated areas and may spread to the new irrigation development areas if proper precautions are not taken.

---

*Jon Hinderink and M. Kiray, "Interdependencies Between Agronomic Development and Social Change: A Comparative Study Conducted in the Cukurova Region of Southern Turkey," Journal of Development Studies (July 1968).

The literacy rate of the adult population of Turkey has increased from about 30 percent to about 50 percent since 1948. However, only about 40 percent of the rural population is literate. Of this group 73 percent is male and 27 percent female. The situation is improving, however, with the school attendance rate of children in the 7-12 age group showing a rise from about 65 percent in 1948 to 94 percent in 1965. Despite this gain critical shortcomings continue to exist in the teaching programs.

Although Turkey has shown rapid development in most social fields, the country continues to fall far short of modern standards in these areas. As indicated in Table 1.4, books published in Turkey increased 100 percent between 1950 and 1964 and newsprint consumption rose 200 percent. At the same time the number of inhabitants per physician is still 10 times higher than it is in developed countries, and care is concentrated in the urban centers at the expense of the rural areas. As late as 1937 Turkey had only 6,700 primary schools. This number has now increased five times, but still it is insufficient.

While the population of Turkey is growing at the rate of about 2.9 percent per annum, the death rate has declined during the last generation. The birth rate has not yet shown any significant tendency toward a decline. The population growth rate is highest in the less developed eastern part of the country. If a line were drawn from Adana to Samsun, it would show that twice as many people live west of the line as east of it, but that about the same number of babies are being born annually on each side of the line. Most of the country's industry would be located on the west side of the line, while the economy on the east side would be predominantly agricultural.

Family planning is beginning to take effect in certain areas and the birth rate is starting to decline in western Turkey. However, a relatively high population growth rate is likely to continue for some time for the country as a whole, since modern public health services are just starting to take effect in eastern Turkey, where the infant mortality rate is still unusually high.

Since 1935 the urban population has been growing faster than the rural population. Yet the rural population has shown a continual growth--from 1965 to 1970 the annual growth rate was about 1.5 percent. The movement of workers to Western Europe began to reach significant dimensions after 1963. The number of Turkish people employed abroad is now estimated at nearly half a million, the majority of them working in West Germany. Employment

TABLE 1.4

Selected Social Statistics for Turkey, 1937-65

	1937	1950	1955	1960	1964	1965
Illiterates among population of age 6 and above (percent)	n.a.	65.3	57.6	56.0	51.7	n.a.
Number of primary schools	6,700	17,428	18,724	24,398	29,363	30,994
Number of students at institutions of higher education	n.a.	25,000	37,192	65,297	85,989	97,308
Books published	n.a.	2,150	3,260	3,447	5,744	n.a.
Newsprint consumption (kilograms per capita)	0.5	0.7	0.7	0.8	1.1	n.a.
Number of radio licenses (1,000)	46	301	999	1,341	2,177	2,443

Note: n.a. = Data not available.

Source: Z. Y. Hershlag, Turkey: The Challenge of Growth (Leiden: E. J. Brill, 1968).

abroad has become a major factor both in the distribution of manpower at home and in terms of its impact on Turkey's balance of payments. It is estimated that overseas nationals generated income of $273 million in 1970 and $471 million in 1971.

Figures on seasonal unemployment in Turkish agriculture for 1967 given in the Second Five-Year Development Plan show that unutilized manpower ran very high for most seasons: about 78 percent of the agricultural labor force from December to February; and more than 50 percent from March to June and from September to November. Unemployment reached a seasonal low of about 9 percent in July and August. These estimates suggest that more than half of the rural labor force is unemployed except for two months each year. The figures also indicate that the landless laborer is likely to be employed only for a few months each year during the busy season.

The proportion of women in the overall labor force in Turkey--nearly 40 percent--seems impressive. However, this figure represents mainly agricultural employment. About 95 percent of female labor is employed in agriculture, with only 3 percent in industry and 2 percent in services. Of the total employment in agriculture in 1960, about 9.7 million, approximately 4.7 million were men and 5.0 million were women.

The demographic and social structure of Turkey has entered a period of far-reaching transformation. The main factors involved are the following:

1. rapid population growth, especially in the less developed rural areas of eastern regions
2. a total cultivated acreage which may have passed its economic limits, initiating a push out of agriculture
3. far-reaching improvements in transportation, triggering and facilitating population movements from the rural areas to major urban centers
4. the pull of the major urban centers
5. the difficulty, with 60 percent of the population in agriculture, of developing other sectors fast enough to absorb the total increase of the working population
6. a considerable amount of migratory labor employed in Western Europe
7. the gradual economic adjustments required for membership in the EEC.

The high population growth in Turkey's less developed areas is not offset by relatively higher public investments in these areas. There are indications that the policy in Turkey is such that investments are directed more toward the developed regions.

The country may face the problem of having to employ additional people in agriculture, thereby creating a decline in acreage of cultivated land per person. This points to the urgent need to increase agricultural productivity per hectare through higher yields, higher-value crops, multiple cropping in the irrigated areas, and more intensive livestock production.

CHAPTER 2

THE NATURAL RESOURCES

GEOGRAPHIC FEATURES

Turkey has an area of 781,000 square kilometers, or more than three times the area of the United Kingdom; and all but 3 percent of Turkey is in Asia. A distinctive feature of the Asiatic portion of Turkey (called Anatolia or Asia Minor) is the high mountains that surround it on all sides, making a basin-like formation and separating it from the sea and from the Asiatic mainland. The central plateau forming the bottom of the basin consists of a steppe with levels of elevation that range in the west from 800 to 1,000 meters and from 1,000 to 1,200 meters, and in the east from 1,600 to 2,000 meters. The central plateau is bordered on the north and on the south by high mountains, mainly between 1,600 and 3,000 meters above sea level.

Turkey has a larger proportion of mountain area than most other countries. Many valleys and narrow coastal plains provide good agricultural regions, but the mountains dominate the scene. The destruction of forests and overgrazing have left much of the soil in an eroded and barren condition. Therefore the characteristic Turkish landscape is one of barren mountains with more fertile valleys in between.

South from the Black Sea the mountains run roughly from east to west. Eastern Turkey is marked by a series of parallel ranges, divided by narrow valleys, extending along the entire coast of the Black Sea, with a number of fast-flowing rivers running between the inner ranges. The Black Sea coastal plain is narrow, widening only near the rivers' estuaries.

Like the Black Sea coast, the Mediterranean coast is cut off from the interior of Turkey by a high mountain range, the Taurus. In the extreme southwestern corner of Turkey the mountains extend to the sea. Further east the mountains turn inland and extend around the fertile Pamphylia plain near the port of Anatolia. East of Anatolia the Taurus mountains again extend close to the sea, separating the Pamphylia plain from the much larger plain of Cukurova.

The land along the western coast of Turkey differs considerably from the northern and southern coastlands. The river valleys which lie between the east-to-west mountain ranges are wider with extensive fertile plains near the coast.

CLIMATE

Turkey's climate is mainly determined by her physical features. The moisture in the winds that come from the Black Sea is precipitated mainly in the form of rainfall on the northern slopes of the mountains running from east to west. These slopes receive the highest annual rainfall in Turkey and are heavily forested. Precipitation in the Black Sea area occurs primarily in the summer months.

The southwesterly winds from the Mediterranean discharge their moisture on the high mountain areas bordering on the central Anatolia plateau. In the west of Turkey rainfall reaches further inland than along the Black Sea because of the wide valleys going from west to east, which cut into the country by about 300 kilometers. The southern and western coastal regions have a Mediterranean-type climate with rain primarily in the winter months.

The central part of the Anatolia steppe region has the lowest annual rate of precipitation in Asia Minor, ranging from 200-600 millimeters (one millimeter is equal to 0.04 inches). The precipitation that does occur falls primarily in the winter and spring months.

Northeastern Turkey is well fed by rainfall; the climate is cold without a dry season. Data concerning the number of months of frost reveal that eastward from Ankara the winter climate gradually becomes more severe. While there is only one month of frost in Ankara there are four in Van and Erzurum.

During the year the average rainfall varies between a maximum of 3,000 millimeters in the eastern parts of the Black Sea coast (Trabzon, Rize) and a minimum of less than 200 millimeters in the area around the salt lake near Konya. Obviously in a country as mountainous as Turkey there are many different local microclimates which vary with the run of the valleys and other topographic features. (See Figure 2, which shows the regional and seasonal distribution of rainfall in Turkey.)

Averages with regard to rainfall in Turkey can be somewhat misleading. Major fluctuations in precipitation are apt to occur from year to year and often several years of drought follow each other. Weather data and studies of the growth rings of trees show that these fluctuations have been typical for many centuries. In a major portion of Turkey rainfall is scarce even in a normal year. Except for the eastern part of the Black Sea coast and some of the mountainous regions, irrigation is essential for achieving high yields and cropping intensity. Without irrigation poor harvests may occur over wide areas when rainfall is below the average.

In summary it can be said that in a major part of the country water is the decisive factor limiting agricultural production. The risk factor, involved in rainfed agriculture in areas with large fluctuations in rainfall, also limits the use of such inputs as fertilizers.

LAND RESOURCES

The soils of Turkey vary greatly in their type. Most of them have good structure with a high pH (7.5 to 8.5) and high lime content. The salt content of the soil is often relatively high and irrigated land is frequently saline. A small portion of the soil along the Black Sea coast is acid, which makes it possible to grow tea and other acid-tolerant plants. The soils are almost uniformly deficient in phosphate and nitrogen. Potash deficiencies are limited to small geographic areas, but may expand with more intensive cultivation and the use of high rates of nitrogen and phosphate applications. Near Samsun cobalt and molybdenum deficiencies have been observed. The soil is generally low in organic matter due to climatic conditions and poor soil management practices. Since the majority of Turkish soils are heavy in texture and high in clay, organic matter is important in improving soil texture. The content of organic matter can be increased through the use of manure. However, it is estimated that 78 percent of the annual manure produced is burned as fuel.

Data in Table 2.1 describe the land resources of Turkey and show the long-term changes in land use patterns. Regarding the figures for the area cropped it should be noted that fallow land is included (the sown area each year is about 35 percent less) and that two categories--"vineyards and olive groves" and "fruits and vegetables"--are not included, but have to be added separately to get the "total cultivated area" of 27.5 million hectares, or 35.5 percent of the total land area. The areas of natural grazing and of forests account for, respectively, 24.5 percent and 23.7 percent of the total land area.

The data show a large increase in cropped area corresponding to a reduction in grazing area. These significant changes in land use are primarily the result of increasing population pressure and growing mechanization. A sharp expansion of the cropped area took place between 1948 and 1953. The introduction of nearly 40,000 tractors and a number of combines, together with the development of the road system, high price supports for wheat, and exceptionally favorable weather in 1951, 1952, and 1953, encouraged farmers to plow up previously uncultivated land. Much of this land is marginal and even submarginal, representing an abortive "wheat revolution" that should have been prevented, at least partly, according to FAO experts. These experts note that the "wheat revolution" represented an overextension of the area of cultivation, an overgrazing of grasslands, and an encroachment of the cultivated and grazing areas on the forests. FAO estimated that the average acreage plowable, without the danger of serious erosion, is only about 16 million hectares or two-thirds of the total acreage. They recommended that parts of the cropped area be used for agricultural purposes only after soil and water conservation practices have been instituted. The other parts are subject to heavy erosion, FAO noted, and should be used for pasture only after the establishment of continuous vegetation.

In 1959 FAO referred to these cultivation problems as a "national emergency" and called for a reduction in nonirrigated arable land and a reduction in the number of livestock in relation to the capacity of the pastures.* A reduction in central Anatolia's livestock to two-thirds of its current number was indicated as desirable.

---

*FAO Mediterranean Development Project, <u>Country Report for Turkey</u> (Rome, 1959).

FIGURE 2

Regional Distribution of Rainfall in Turkey, 1950

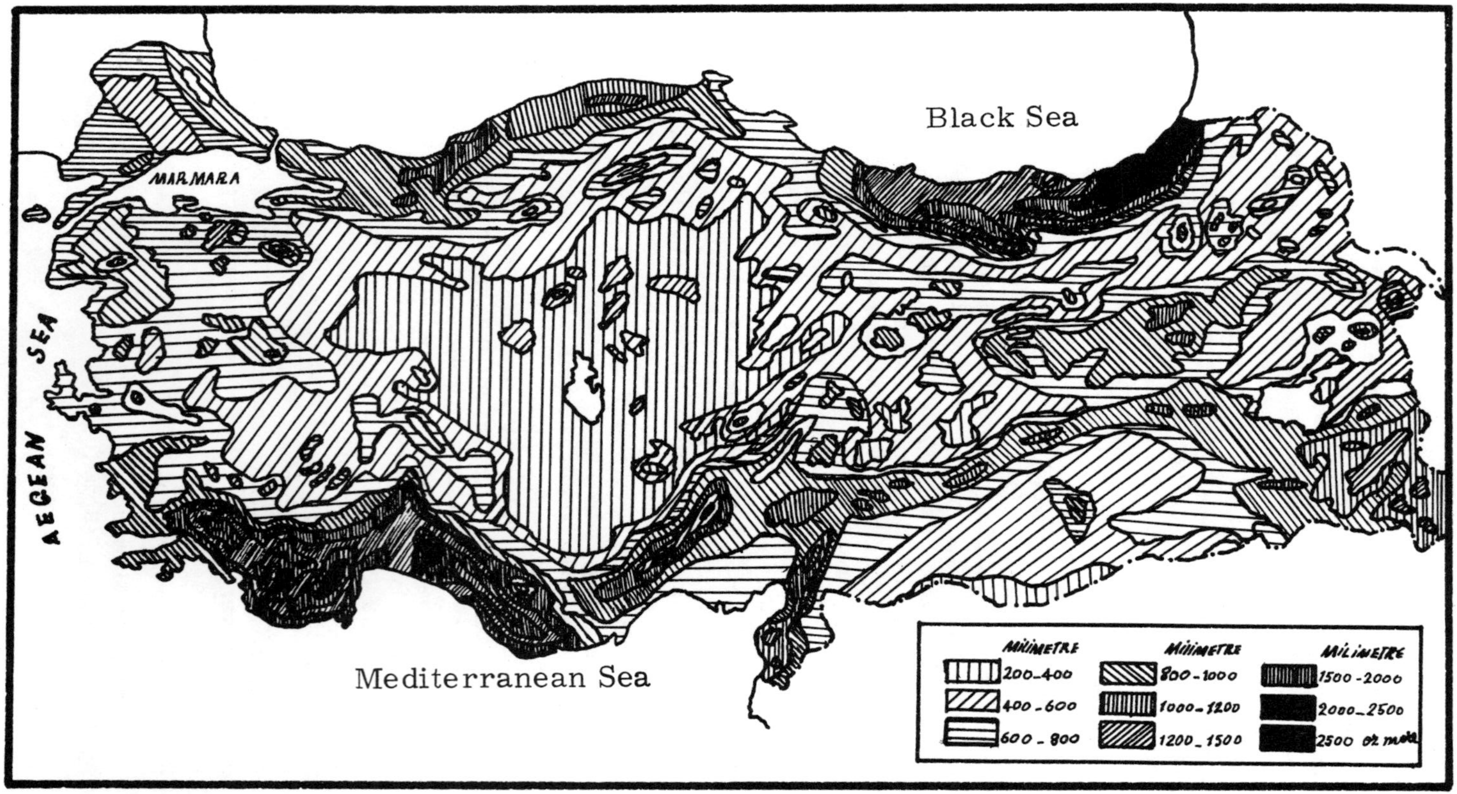

Seasonal Distribution of Rainfall in Turkey, 1950

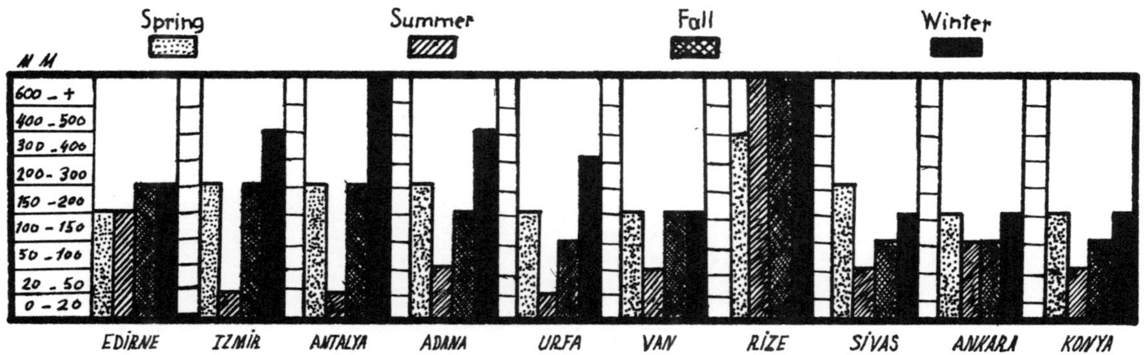

Source: Mediterranean Development Project Country Report--Turkey.

TABLE 2.1

Patterns of Land Use in Turkey, 1927-71
(1,000 hectares)

Year	Total Cropped Area	Percent Fallow of Cropped Area	Grazing Area	Vineyards and Olive Groves	Fruits and Vegetables	Forests
1927	7,600	n.a.	46,400	n.a.	n.a.	n.a.
1938	13,158	35.6	41,000	660	650	10,386
1946	13,093	36.0	n.a.	n.a.	n.a.	n.a.
1952	17,413	36.0	34,789	1,031	636	10,418
1956	22,453	35.1	29,489	1,195	682	10,418
1961	23,076	34.4	28,767	1,361	778	10,584
1966	23,982	35.6	28,013	1,496	888[a]	10,584
1967	22,896	34.9	26,135[b]	1,514	900	12,578
1968	24,092	36.0	19,622	1,571	1,405	18,273
1969	24,731	35.2	19,051	1,561	1,451	18,273
1970	24,294	35.8	19,313	1,576	1,467	18,273
1971	24,424	35.2	n.a.	1,587	1,500	18,273
1972	25,073	36.0	n.a.	1,601	1,582	18,273

[a]Including tea.

[b]Reductions in 1967 and 1968 are due to changes in classification of land as forest.

Notes: Small anomalies occur in the figures due to a revision of the total area of Turkey. Land use statistics have been revised in the light of the recent discovery that the total forest area, formerly believed to be 10.5 million hectares, actually exceeds 18 million hectares. n.a. = Data not available.

Effective programs to implement the FAO proposals have not emerged. However, the rate of expansion of cropping into marginal pasture land, and the resulting reduction of the grazing area, appear to have come to an end, though the number of animals is still increasing.

It seems obvious that only marginal land is left uncultivated in the dry, rainfed areas of Turkey, and that no further economical expansion of cropping is possible in these areas. The expansion of irrigation facilities will serve mainly to increase the productivity of existing arable land.

According to Necmi Sonmetz, there are some areas in which cultivation can be expanded by the reclamation of saline-alkaline soils and swamps and by improvement of strips of land in coastal areas.* It is estimated that approximately 3.3 million hectares could be reclaimed in this way. In addition there are about two million hectares of land which can be cultivated by reclamation of wild olive groves, pistachio forests, and other bushes.

An important development during the last decade has been the doubling of the area producing fruits and vegetables. This development toward high value crops is likely to continue in the most favorable coastal areas, especially where a perennial supply of irrigation water is made available.

For the country as a whole the percentage of fallow land seems to have remained at the same level. However, the percentage of fallow land in the dry and semidry regions rose from 31.8 percent in 1948 to 35.8 percent in 1970. This can be explained by the increasing erosion of the land and by the expansion of farming into relatively more and more marginal land, which should have been preserved as pastures. The percentage of fallow varies from one region to another: in south and east central Anatolia, where the climate and land conditions are poor, it is 42 percent; in southeastern Anatolia it is as high as 45 percent; and it is between 17 and 20 percent in the Marmara and Black Sea regions.

---

*Necmi Sonmetz, "The Existing Land Ownership and Land Use Conditions in Turkey" (University of Ankara: Yearbook of the Faculty of Agriculture, 1963).

WATER RESOURCES

Irrigation Potential

In view of Turkey's relatively dry climate, the availability of irrigation could become a very important factor in the country's efforts to develop a highly productive, diversified, and intensive agriculture. Irrigation is, more or less, essential in all parts of the country. With efficient irrigation as a major part of a highly adaptive package of new technology, production per hectare could show manifold increases.

Although the first modern irrigation system in Turkey, located at Konya, dates back to 1905, the potential of water resources and the area suitable for irrigation is still under investigation. Of the total average annual amount of precipitation, 167,000 million cubic meters (135.4 million acre-feet) is lost in the form of runoff. The entire runoff is unusable in Turkey, because of the water rights of neighboring countries and the need for dilution and disposal of the polluted waters. In addition either the topography and geological conditions in some basins are not suitable for the construction of flow-regulating dams, or where dams can be built it is not possible to fully utilize the water captured in the same basins. The amount of runoff that can be utilized is estimated to be 155,000 million cubic meters (126 million acre-feet).

Extensive groundwater investigations have been carried out in Turkey during the past 10 to 12 years. In the areas where such investigations have been completed, the dependable water potential that can be utilized is about 4,000 million cubic meters (3.24 million acre-feet). However, it is estimated that for the entire country groundwater resources approximate 6,000 million cubic meters (4.86 million acre-feet).

Before 1966 the use of groundwater for irrigation was very limited. But in 1966 three state agencies--the General Directorate of the State Hydraulic Works, Deolet Su Isleri (DSI), the Soil and Water Directorate (Toprak Su), and the Agricultural Bank of Turkey--signed a protocol to accelerate the development of underground water resources for irrigation.

It has been estimated that the total area physically suitable for irrigation is about 12.6 million hectares and that there are sufficient surface water resources to irrigate even more land. From this point of view the amount of irrigable land for which the water can be profitably used is the limiting factor, not the water. Earlier estimates showed that less than half of the physically suitable area (about 6 million hectares) can be irrigated at an economically acceptable cost--that is, without constructing expensive pumping structures and without transporting the water from one watershed to another.

In the Second Five-Year Plan it is indicated that the potential exists to irrigate 7.0-8.5 million hectares of land. The extent to which water resources will be found economical to develop depends of course upon many factors, including efficiency in using the water, the yield levels obtained, product prices, and engineering technology. However, through the application of new technology, it may ultimately become economical to exploit a proportion of the total water resources, which would have the capacity to irrigate nearly one-third of the cultivated area of the country. The production value of this acreage would most likely add up to 75 percent or more of the total production value of agriculture.

According to a recent FAO document,* the corresponding figures for the Near East (excluding Turkey and Pakistan) are about 20 percent of total cultivated acreage and 74 percent of total agricultural production value. The most valuable crops, such as fruits and vegetables, will be far more dominant on irrigated land than on non-irrigated land. From a national economic viewpoint large investments for rapid development of water resources may therefore be justified. In the long run the irrigated crops located mainly in the coastal areas and the south-east region, may dominate Turkish agriculture. This would mean that the central regions would lose ground in relative agricultural importance.

The area of agricultural land now under irrigation is estimated to be about 1.8 million hectares. It can be subdivided into the following categories:

	Hectares
Private irrigation, unimproved	806,000
Private irrigation reconstructed by state agencies	194,000
Government irrigation programs	800,000
Total	1,800,000

*Prospects of increasing production of rainfed agriculture with particular emphasis on low rainfed areas. Paper prepared for the Twelfth FAO Regional Conference for the Near East, Amman, Jordan, 31 August-9 September 1974.

This tabulation does not give a complete picture, since in recent years little more than half of the area commanded by government works has received material benefit. In 1967 the area that was actually irrigated by government programs amounted to only 345,100 hectares. For the remaining area under government development various factors have prevented effective distribution of water--approximately 60 percent of the water in the state systems is obtained from storage dams, 36 percent is obtained from diversion dams, and 4 percent from the underground by tubewells.

As a result of the variations in climate and soil characteristics, the kinds of crops grown on irrigated lands show great variations from region to region. Tables 2.2 and 2.3 show the variety of crops grown in Turkey, giving the amount and percentage of irrigated area for each crop.

The tables indicate that cotton, in relation to the other crops, has been occupying a growing portion of the irrigated land and is now grown in nearly half of the area. In the Aegean region it is grown in nearly 80 percent of the irrigated area. For wheat the figures indicate a relatively declining share of the irrigated land from 1962 to 1967, followed by an increase, which resulted from the introduction of the Mexican wheat varieties in the coastal regions. However, recent reports indicate that wheat again has started to become less and less important on the irrigated land.

Efficiency of Water Use

Although investment in irrigation works is consuming about 30 percent of the government's total investment budget and about 70 percent of investment in the agricultural sector, both the Turkey Country Report of 1959 by FAO and a report on prospects for Turkish agriculture of 1966 by a USAID study team headed by Dr. C. M. Elkinton paint a dark picture of the efficiency of the distribution and the use of water in Turkey. FAO emphasizes the following difficulties (in its 1959 Country Report for Turkey):

1. lack of coordination in the construction programs for dams and in distribution and drainage systems
2. general lack of knowledge on the part of the farmers regarding modern irrigation methods and practices
3. lack of suitable farm machinery and equipment for irrigation farming

4. lack of a suitable technology package to go with the water
5. inadequate drainage leading to waterlogging and salinity
6. introduction of no new suitable cropping patterns.

TABLE 2.2

Distribution of Irrigated Crops in Turkey, 1962 and 1967

	1962		1967	
	Irrigated Area (in 1,000 hectares)	Percentage of Total Irrigated Area	Irrigated Area (in 1,000 hectares)	Percentage of Total Irrigated Area
Wheat	200	17.9	200	12.9
Cotton	250	22.4	400	25.8
Vegetables	200	16.1	240	15.5
Citrus fruit	25	2.2	37	2.4
Other fruit	145	14.8	140	9.0
Maize	70	6.3	75	4.8
Sugar beet	82	7.3	110	7.1
Rice	60	5.4	57	3.7
Feed and forage crops	30	2.7	n.a.	n.a.
Others	53	4.7	n.a.	n.a.
Other grains	n.a.	n.a.	40	2.6
Legumes and fodder	n.a.	n.a.	100	6.5
Other industrial crops	n.a.	n.a.	80	5.2
Sunflower seeds	n.a.	n.a.	70	4.5
Total	1,115	100.0	1,549	100.0

Notes: Some of the areas are double-cropped each year. This explains the apparent inconsistency between area figures and percentages in the table; n.a. = data not available.

Source: First Five-Year Plan and Second Five-Year Plan, Turkey.

TABLE 2.3

Distribution of Irrigated Crops in Turkey, 1970

Crop	Percentage of Total Irrigated Area
Cotton	48.52
Wheat	19.26
Fruits	6.62
Sugar beet	4.55
Vegetables	4.31
Forage crops	3.60
Corn	3.11
Watermelon	2.94
Pulses (leguminosae)	2.39
Oil crops	2.19
Citrus fruit and bananas	1.97
Tobacco	0.75
Potatoes	0.46
Poplar trees	0.33
Total	100.0

Source: Kaya Bozkurt, "Irrigation and Soil Improvement Problems in Agriculture" (Istanbul: Economic and Social Studies Conference Board, October 1972).

The AID study team (in its report issued in 1966) points to the following problems:

1. the lack of land leveling that is essential for efficient water management
2. wild flooding in applying water that results in very low water use efficiency
3. lack of digging equipment to install well tubes.

Because of these problems only 50 percent, or less than 50 percent, of the expected advantages of the irrigation systems have been obtained. The deficiencies seem to derive in large measure from a marked imbalance in the purposes and programs of the three major water development agencies: DSI, Toprak Su, and the Extension Service of the Ministry of Agriculture. DSI is responsible for the design and construction of major irrigation works, while Toprak Su is responsible for land preparation and all on-farm development works. In addition Toprak Su is charged with the responsibility for the design and construction of small irrigation projects (having a discharge of less than 500 liters per second).

Two separate and potentially conflicting strategies have recently been used for on-farm water development in Turkey. One plan is used within state project areas--most notably Seyhan and Gediz--and features government-financed, government-executed on-farm development. The other plan is used primarily outside state project areas (in the Izmir area) and centers around farmer-financed, private contractor-executed on-farm development. In this case Toprak Su provides only technical assistance to farmers planning the improvement of their land; the farmers then finance the actual land improvement. As a result of the high returns on investments, demand for land leveling has soared in the Izmir region.

While the government-financed strategy has proved effective in bringing rapid on-farm development to selected projects, this approach irrevocably links the potential development to the size of the Toprak Su budget. While an increase in the size of that budget from 21 percent to one-third of the total agricultural investment budget might permit Toprak Su's equipment supply to be more than doubled, the backlog of 400,000 hectares of undeveloped land in state project areas would continue to grow. Moreover this approach does nothing toward the development of the approximately one million hectares of irrigated land outside state project areas, which is nearly twice the land that is within state areas. Since the private-investment strategy is financed not by general budget revenues, but by the increased income from development, there is no ceiling on development--as there is under the first approach.

A comparison of the two strategies reveals the following key differences:

1. Returns on scarce government funds (as opposed to total investment) are far higher under the "private" strategy.

2. The "private" approach creates more opportunities for both employment and entrepreneurship than does the government approach.

3. The "private" strategy does not increase income inequality as sharply as does the government approach since in Izmir the beneficiaries of development pay the cost. In the period from 1966 to 1970 half of the

leveling in Seyhan (Adana province) went to 152 farmers. This government investment has added, on the average, about TL 766 thousand to each farmer's annual income. For the 23 largest farmers, government investment has increased annual average income per farmer by over TL two million.*

It appears that through the privage-investment strategy, which mobilizes private capital to pay for on-farm investment, all the elements are present in Turkey for a truly spectacular increase in the productivity of irrigated agriculture.

Present Irrigation Programs

The total area for which irrigation programs are under consideration or in various stages of implementation amounts to about 2.4 million hectares. If this entire area were successfully developed, and high-yield crops and varieties were to result, it would obviously make a great impact on the total level of agricultural production in Turkey.

Improvements in land leveling, drainage, and water management would be necessary to obtain higher profits in irrigated agriculture and to assure higher returns from the investments in irrigation. A survey of 18 state irrigation systems conducted in 1965, showed that 53 percent of the land required a light degree of leveling, 29 percent required a medium degree, and 8 percent a heavy degree. In addition it was found that 40 percent of the land required surface drainage and 26 percent required tile drainage; also that 73 percent of the land was being irrigated through the method of wild flooding, 17 percent through the basin method, and only 10 percent through the furrow method, which is the more efficient one.

Irrigation is considered to be one of the most important factors in producing higher crop yields. The various factors accounting for increased total agricultural production in Turkey, as accepted by the State Planning Organization, are listed as follows, along with the percentage of increased output accounted for by each factor:

---

*Charles K. Mann, "Formulating a Consistent Strategy Toward On-farm Land Development in Turkey" (Istanbul: Economic and Social Studies Conference Board, October 1972).

Irrigation and increased water use efficiency	20
Fertilizers and good-quality seeds of high-yield varieties	62
Decrease in fallowing	8
Weed control and pesticides	10
	100

Various factors influencing production on irrigated land, observed by Toprak Su in the Aegean region after the establishment of land development services, are listed as follows, along with the percentage of increased output accounted for by each factor:

Land leveling and improved irrigation methods	40
Good-quality seeds of high-yield varieties, fertilizers, and crop rotation	40
Modern tools and equipment	10
Weed control and pesticides	10
	100

CHAPTER

3

LAND USE CHARACTERISTICS

NINE LAND USE ZONES

Starting with the agricultural census of 1950, Turkey has been divided into nine agricultural regions or land use zones, as illustrated in Figure 3. In the following discussion of these zones the coastal regions will be considered first, followed by the central and eastern regions. The three central regions (comprising Central Anatolia) will be discussed jointly.

The Black Sea region (no. VII) extends from the coast to the mountains near the Black Sea. The eastern section of this region has the highest rainfall in Turkey. The cultivated portion of the coastal strip produces hazelnuts, which represent one of the most important Turkish export crops, and tea. The western portion of the Black Sea region, from Ordu to Zonguldak, has considerably less rainfall. The coastal plain formed by the delta of the Kizilirmak and Yesilirmak rivers is very fertile. This plain (Schars Schamba) is one of the most important areas in the country for the cultivation of maize, which is the main food grain in this part of Turkey. This region accounts for only about 5 percent of the total wheat production of Turkey. Tobacco is grown in the dryer and less fertile areas. The number of cattle per square kilometer is high especially along the Black Sea coast. Investment for the development of a dairy plant near Samsun is included in the Second Five-Year Plan.

The Marmara region (no. III), around the Marmara Sea, consists mainly of plains and plateaus, that do not exceed 200 meters in height. The climate is moderate with rainfall occurring mostly in winter. Irrigation would

FIGURE 3

Nine Agricultural Regions, Turkey

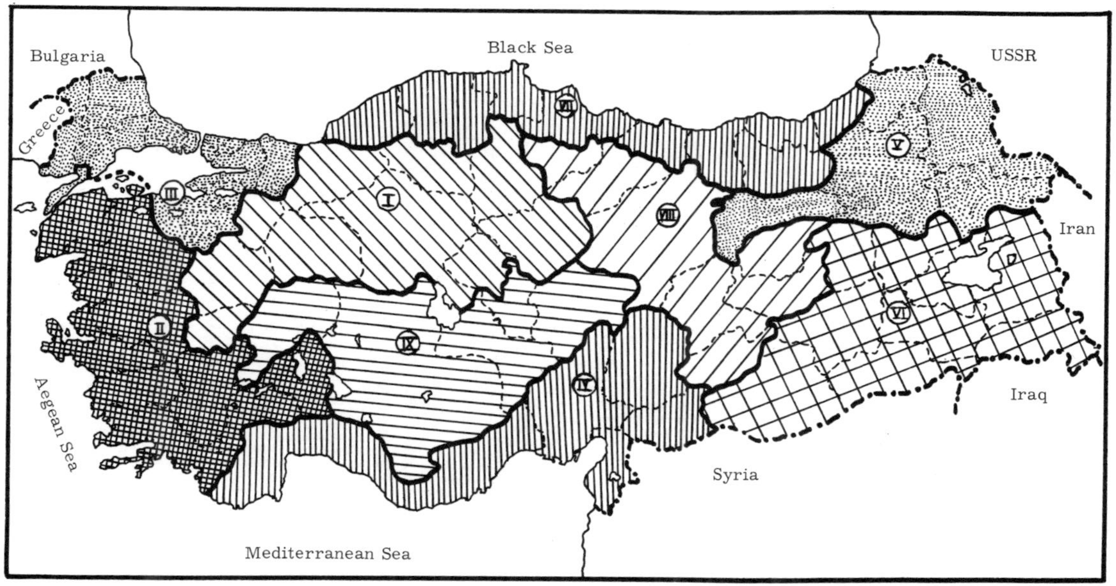

I	Central North	IV	Mediterranean	VII	Black Sea
II	Aegean	V	North East	VIII	Central East
III	Marmara	VI	South East	IX	Central South

<u>Source</u>: State Institute of Statistics.

therefore be important. Wheat, maize, sugar beet, tobacco, sunflower seeds, olives, and livestock are the main agricultural products. The region accounts for 8 percent of the total wheat production. The demands of the Istanbul market have encouraged the growing of citrus fruits, vegetables, and grapes. Extensive cultivation of poplar trees has been started on the plain along the Gulf of Ismit.

The Aegean region (no. II), with fertile coastal plains and river valleys, has a virtually dry summer. Wheat, barley, olives, figs, and grapes grow in this region without irrigation, while cotton and maize, as well as many horticultural products, require irrigation facilities. Cotton production has been expanding. Cotton is now grown on 80 percent of the total irrigated area of approximately 440,000 hectares. This region has a high potential capacity for exports of fresh fruits and vegetables. Cattle and sheep are most important in the northern part of the region. The wheat in the region is planted in October and November, and harvested in June. The region accounts for 10 percent of total Turkish wheat production. The provinces of Manisa, Izmir, Aydin, and Dinizli represent an area of wide irrigation potential. They include the Gedez and Buyuk Menderes irrigation projects, and comprise a total irrigable area of about 245,000 hectares. The landholding patterns in the area are favorable for agricultural development.

The Mediterranean, or southern region (no. IV) covers all of Turkey's southern coast up to the Gulf of Iskenderun. It consists mainly of a narrow coastal strip except for some fertile plains. For agricultural purposes two plains are of particular importance--the plain of Antalya and the plain of Cucurova, in Adana province, on which the Seyhan and Ceyhan rivers have formed a huge delta. The plain of Antalya has the most markedly subtropical climate of the region, with the average temperature in the coldest month (February) ranging from 10°-11° centigrade. Frosts are unknown in normal years. Rainfall is inadequate for intensive cultivation or high cropping intensity. This area has high priority for irrigation, since only a small part of the plain is presently irrigated. Five rivers that cross the plain are well located for multipurpose projects. About 60,000 hectares can be potentially irrigated in this coastal zone. However, the high winter rainfall and runoff from the surrounding mountains give rise to a flooding problem. The landholding pattern is unfavorable. This area has a great potential for subtropical fruits, such as citruses and bananas, and for winter vegetables. On the Cucurova plain cotton and wheat are predominant. Except for some particularly sheltered coastal strips the climatic conditions here are less favorable for the production of early fruits and vegetables than on the Anatalya plain. The Cehan-Aslantas irrigation project spans the provinces of Adana, Icel, and Hatay. The irrigation potential of this project totals about 300,000 hectares and the landholding patterns in the area are favorable for agricultural development. Other crops grown in the region are wheat, maize, sesame, olives, dates, groundnuts, and rice. The region accounts for 15 percent of total Turkish wheat production and Adana province alone accounts for 6 percent. Goats, cattle, and dairy products also play a relatively important role on the Cucurova plain.

<u>The three regions of central Anatolia--Central North, Central East, and Central South</u> (nos. I, VII, and IX) differ from the coastal regions in that they have low rainfall and more severe conditions for agricultural production. Wheat is the main crop, with the Anatolian plateau producing a major part of Turkey's wheat through a rotation system of one year of wheat production and one year of fallow. The wheat is usually planted in October and harvested in July. In several high areas the wheat is planted in March and April, and harvested in July and August. The Central North region accounts for 24 percent of the total Turkish wheat production and the Central South region for 22 percent, while the Central East region accounts for only 5 percent. In the more sheltered parts of the plateau cotton, sugar beets, and tobacco are grown.

The number of cattle, sheep, and goats is relatively high in most of the central provinces. The arable land in this area cannot be expanded much further, if at all. The prospects for irrigation are less favorable than in the coastal regions. The population density is high and the farms are usually small. Because of soil erosion and overgrazing some of the marginal land may have to be taken out of production and the number of animals reduced. The encouraging aspects of this area are the prospects for (1) increasing wheat yields through new varieties and improved cultural practices, and (2) partly replacing the traditional rotation of grain cultivation and fallow with a rotation of grain and fodder legumes, especially in the areas with slightly higher rainfall. Research results indicate that with the type of fallow now used, no significant amount of water is stored from one year to the next. The main effect of the fallow seems to be an accumulation of nitrogen and possibly phosphorus, which could be supplied through artificial fertilizers. On the arable land mainly grain is being cultivated. There is practically no cultivation of forage plants, and potatoes are grown in small gardens. The cattle are mainly fed on straw in the winter. Therefore only cattle breeds fit to endure long famines are kept, and these have a very low yield capacity.

The northeast region (no. V) is formed mainly of valleys, at elevations of 900-1,000 meters, surrounded by a high plateau of 1,200 to 2,000 meters. The region is mountainous and the climate is severe. Winter may last six to seven months, during which time 25-40 percent of the precipitation occurs. The principal crop is summer wheat and the region accounts for about 3 percent of total Turkish wheat production. Some summer barley is also grown. The summer wheat is planted mainly in April and harvested in August. Fruit trees such as apricots are grown in the sheltered areas. Many of the provinces of this region have large numbers of cattle and sheep. The region also has natural prairies for sheep and cattle grazing. A modern dairy industry is being developed in the best cattle areas.

The southeast region (no. VI) slopes downward from the Taurus mountains to a level of 400 meters above sea level north of the plains of Syria. The area is in part a steppe characterized by a nomadic people, the Kurds, who form a large part of the local population but are not yet fully settled. Thus this part of the country has been an unruly frontier. The region is crossed by the Euphrates and Tigris rivers. A potentially large irrigation program, the Lower Euphrates project, might be of great importance in the future. The chief crops are winter cereals, wheat, and barley. Wheat is planted in October and November, and harvested in June and July. The region accounts for 8 percent of total Turkish wheat production. Parts of the region, such as the provinces of Urfa and Van, have many sheep, goats, and cattle, and dairy processing plants have recently been erected.

FARM SIZE AND LAND TENURE

Most of Turkey's rural population live on small family farms. There is a total of about 3.1 million farms in Turkey, with an average farm size of nearly 9 hectares (including the area cultivated, vineyards and olive groves, and fruit and vegetable acreage). Population pressure and traditional rules of inheritance are causing the subdivision of large farms to proceed rapidly. Table 3.1 gives the Turkish land distribution for 1963 and for 1970.

TABLE 3.1

Turkish Land Distribution--Number and Size of Agricultural Holdings, 1963 and 1970

Size of Holdings (hectares)	Holdings within Each Group: Number 1963	Number 1970	Percent 1963	Percent 1970	Percent of Total Land Area within Each Group 1963	1970
0.1-5.0	2,132,291	2,316,243	68.6	75.0	25.0	31.3
5.1-20.0	853,425	673,121	27.6	21.9	42.0	47.5
Over 20.0	115,087	91,302	3.8	3.1	33.0	21.2
Total	2,100,803	3,080,666	100.0	100.0	100.0	100.0

Note: The 1970 results are preliminary. The data are for operation, not ownership, of property. In 1963 a reputed 308,000 rural families having no land, or less than one decare of land, were excluded.

Source: Data derived from Turkish Agricultural Census, 1963, and Turkish Agricultural Census, 1970.

The percentage of small holdings (less than 5 hectares of land) has been increasing during the two last decades. In 1952 the percentage of these holdings was 62.1 and the land cultivated by them was 18.6 percent of the total arable land. These figures rose in 1963 to 68.6 percent and 25.0 percent, respectively, and in 1970 to 75.0 percent and 31.3 percent. The overall figures indicate that the number of small agricultural holdings increases while medium-size and large-size holdings are split into smaller ones. While in 1952 the percentage of large holdings commanding over 50 hectares was 1.5 percent, with their cultivated area amounting to 24.8 percent, these percentages declined, respectively, to 0.5 percent and 13.2 percent in 1963, and to 0.4 percent and 6.0 percent in 1970. The large holdings are being divided rapidly and the average size of the large holdings is diminishing even faster than their number. In recent years there has been a slight increase in the average size of the holdings between 5.1 and 20.0 hectares.

Exact data concerning land distribution according to ownership are not available. Such data will not likely be available before the completion of cadastral surveys, which will take decades. Most of the land in Turkey is not registered at land offices. This situation gives rise to disputes, and makes land consolidation and the granting of agricultural credit difficult. Sönmez has estimated that in 1963 about 1.6 million hectares of cultivated land was owned by the state, and was either used as state farms or rented to farmers (see Chapter 2).

Data given in Table 3.2 show the distribution of holdings in 1970 according to the land tenure system. The data indicate that only a relatively limited number of farmers do not own a part of the land that they farm. For 1950 the highest percentage of owner-cultivators (about 83 percent) was reported for the Black Sea area, and the lowest (about 65 percent) was reported for the other coastal regions. Sharecropping was most common in the southeast region and the Mediterranean region. The percentage of the farmers cultivating only their own land rose, for the whole country, from 72.6 percent in 1950 to 83.5 percent in 1970. For a large part of the agricultural area, either land records and cadastral maps are nonexistent or the records are wrong or incomplete.

In the settled areas pasture land is usually jointly owned by the farmers of a village. This encourages the livestock owners to take the great advantage of the pastures without feeling any responsibility for their maintenance, protection, or improvement. Due to the increased agricultural population much of the original pasture land has been opened to cultivation by individual farmers. On marginal land this has often led to the erosion and devastation of land. According to the figures published by the State Institute of Statistics, during the 30 years from 1938 to 1967 a decrease of 13.0 million hectares of pasture land has taken place. After subtracting the 10.8 million hectares of increased farmland, garden, and orchard areas registered during the same period, there is a balance unaccounted for of 2.2 million hectares. This nearly corresponds to the increase in "unproductive lands," which are not considered good for any kind of production. This development may have very serious consequences for the future capacity of livestock production.

TABLE 3.2

Distribution of Turkish Holdings by Land Tenure System, 1970

Type of Tenure	Number of Holdings	Percent of Holdings
Operate only their own land	2,571,910	83.49
Own more than 50 percent of the land they operate	250,916	8.17
Own less than 50 percent of the land they operate	116,643	3.79
Only leasehold	47,100	1.53
Only sharecrop	77,543	2.52
Have only other arrangements	10,116	0.33
Have two or more types of arrangements combined	4,772	0.15
Lease their lands by any arrangement	1,666	0.05
Total	3,080,666	100.0

Source: Turkish Agricultural Census, 1970.

Land fragmentation is widespread because of both the increase in agricultural population and the traditional Turkish rules governing inheritance (see Table 3.3). In 1950 nearly 95 percent of the landholdings were fragmented to some degree, and nearly half were split into six parcels or more. This situation added to the difficulties of implementing irrigation, mechanization, and disease and pest control programs, and resulted in needless waste of time and legal conflicts. The situation has improved considerably during the last two decades. The percentage of single-unit holdings was approximately 5 percent in 1950, 10 percent in 1963, and 22 percent in 1970. The percentage of holdings with more than six units fell in the same period from nearly 50 percent to 29 percent. In 1970 the average number of plots of land per holding was five.

The pattern of ownership and land fragmentation varies both geographically and with the size of holdings. In some areas either there are few economically viable holdings left or ownership of much of the land is in dispute. In other areas the situation is reasonably good regarding both farm size and fragmentation.

TABLE 3.3

Plots of Land in Turkish Agricultural Operations

Number of Plots	Percentage of Total Operations 1950	1963	1970
1	5.4	9.6	21.6
2-3	22.7	20.8	29.7
4-5	23.2	19.9	20.1
6-9	26.1	24.9	17.8
10 and over	22.6	24.8	10.8

Source: Compiled by State Institute of Statistics, Turkey, based on Turkish Agricultural Census for 1950, 1963, and 1970. (The results from the 1970 census are preliminary.)

"Dwarf" farms (0.1-2 hectares) predominate in the northeast and Black Sea regions, where between 44 and 49 percent of the farms are in this category. The Aegean region and the Central South and Central East regions have between 31 and 36 percent of the farms in this group.

The number of farms of 10 hectares or more are most common in the Central South region, where about 28 percent are in this group. The Central North region and the southeast region have about 25 percent of the farms in this group. In the Black Sea area only 5 percent of the farms are in this group. Large farms above 50 hectares are most common in the Mediterranean region and the Central South and southeast regions.

The 79 percent of farms under ten hectares in the Mediterranean region represent only 21 percent of the region's cultivated area, and the 72 percent of farms of this size in the Central South region represent 26 percent of the cultivated area. On the other hand, in the Black Sea region the 95 percent of farms with less than ten hectares occupied 79 percent of the cultivated area.

For the country as a whole about 34 percent of the farmers are engaged only in crop production. This percentage varies from 44-45 percent in the Marmara and Aegean regions down to 18-19 percent in the northeast and southeast regions.

Farmers engaged only in livestock production represented 1.7 percent of the farmers in the country in 1950. This percentage varies from 4.5 percent in the southeast region to 0.3 percent in the Black Sea region. Mixed farming is the predominant type of farming in most parts of the country.

Rural nonfarm families and landless labor represented in 1950 about 15 percent of the rural families in the country. The proportion varied from 22 percent in the Mediterranean and southeast regions to 8 percent in the Black Sea region. Precise data concerning the proportion of the populace categorized as landless labor are currently unavailable. However, estimates suggest that the absolute and relative numbers of landless people have increased, and that this group may represent 20-25 percent of the current Turkish rural labor force.

CHAPTER

4

AGRICULTURAL PRODUCTION AND INCOME

COMPOSITION OF OVERALL AGRICULTURAL PRODUCTION IN TURKEY

In Turkey plant products account for nearly 68 percent of agriculture's total contribution to the national income, while animal products account for approximately 27 percent, as shown in Table 4.1.

The relative role of plant products has been increasing over the last three decades--they accounted for 54 percent of agriculture's contribution to the national income in 1938, 57.6 percent in 1948, 61.3 percent in 1958, and 67.8 percent in 1970. On the other hand the percentages accounted for by livestock products declined from 39.2 percent in 1938 to 37.0 percent in 1948, to 33.6 percent in 1958, and to 26.8 percent in 1970. The growth of animal products has remained far behind that of plant products, and there may be many reasons for this. Because of the increased population a large part of the best village pastureland has been cultivated for wheat production, which again has increased the degree of overgrazing and the rate of deterioration of the grazing lands. The technological improvements seem to have favored plant products in Turkey, while an attachment to traditional methods is still prevalent in animal production. On the other hand this situation indicates that a big technological leap forward could be forthcoming in animal production.

The percentages of agricultural income contributed by agricultural crafts, forestry, and fishing are all rather small, and these figures have had a tendency to decline over the years.

Among plant products, farm products account for nearly 50 percent of agricultural income, while horticultural products account for 18 percent. However, the relative importance of the horticultural group has been increasing. Expressed as a percentage of Turkey's total cultivated area, the share held by this group--vegetable gardens and fruit orchards--increased from 1.7 percent in 1938 to 1.9 percent in 1948, to 2.5 percent in 1958, and to 3.8 percent in 1970. For farm products the respective figures were 17.3 percent in 1938, 17.9 percent in 1948, 29.3 percent in 1958, and 30.9 percent in 1970. The area cultivated in both types of production has increased, but fruit-vegetable production has grown more rapidly, especially in recent years. Compared with the acreage in 1950, the farm products area has expanded approximately 60 percent and the fruit-vegetable area has nearly doubled. In this period the expansion in farm products took place during the first decade, while the growth in fruits and vegetables was accelerated during the last decade. Since these products, especially fruits, have a higher monetary value than most of the farm products, their accelerated growth represents a significant development.

TABLE 4.1

Composition of Agricultural Production in Turkey According to Contribution to National Income, 1970

Subsectors and Their Divisions	Percent of Agriculture's Contribution to National Income
Farming	98.0
Plant products	67.8
Farm products	49.7
Grain	30.3
Pulse	1.6
Other	17.8
Horticultural products	18.1
Animal products	26.8
Agricultural crafts	3.4
Forestry	1.3
Fishing and hunting	0.7

Source: State Institute of Statistics, Turkey--national income figures.

Regarding the total value of agricultural production, the share held by fruits and vegetables increased from 10.8 percent in 1952 to 12.4 percent in 1963, and to 21.4 percent in 1970. This uptrend in the fruit and vegetable area most likely will continue as a main goal of the strategy of agricultural development in Turkey.

In 1970 the total sown area in Turkey had the following composition: Grains accounted for 84.9 percent with wheat alone having a share of 55.2 percent; pulses accounted for 3.4 percent; industrial crops for 6.5 percent; oil seeds for 3.7 percent; and root crops for 1.5 percent. The figures illustrate the dominating role of wheat and other small grains in Turkey, especially when one considers the fact that wheat and barley are grown in rotation with a major part of the 8.5 million hectares of Turkish fallow land. In addition the relative role of wheat has been increasing, since it covered only 45.2 percent of the total area sown in 1951. The urge to feed a rapidly growing population with a high wheat consumption per capita, during a period showing no significant increases in wheat yields, may represent a major reason for this development.

In terms of production value the picture is different. In 1970 the shares in the total value were 26.3 percent for grains, including 17.0 percent for wheat, 16.7 percent for industrial crops, and 1.8 percent for pulses. During the same period 1951-70, while the relative share of total production area held by wheat increased by 20 percent, its relative share of production value in the period 1951 to 1970 declined by 20 percent, illustrating the growing importance of high-value industrial crops. These crops now have a share of nearly 20 percent of the total production value in agriculture in spite of the fact that their relative share of the production area has been decreasing.

The pulses represent a small portion of the production area and an even smaller portion of the value of production. Their relative share of the cultivated area declined from 4.1 percent in 1951 to 3.4 percent in 1970. With the growing shortage of animal protein in Turkey this trend may not continue.

Regarding animal production in 1970, the total number of cattle, sheep, and goats was 73 million, with cattle accounting for 20 percent of this number, sheep for 50 percent, and goats for 30 percent. By contrast the total number of animals in 1952 was 60 million. The increase over the last 20 years has been 25 percent for cattle and

47 percent for sheep, while the number of goats has declined by 10 percent. In 1970 the value of animal products accounted for 31.3 percent of the total agricultural gross production value. The share of meat was 15.4 percent, of milk 12.8 percent, and of other products 3.1 percent. In 1952 the animal products accounted for 35.2 percent, and the respective shares were 14.3 percent, 17.6 percent, and 3.3 percent, indicating the growing relative importance of meat. It should be noted that during the entire period cattle meat has represented only approximately 33 percent of the value of meat. This illustrates the relatively high importance of sheep meat in Turkey.

COMPOSITION OF PRODUCTION IN VARIOUS REGIONS

The regional composition of agricultural production in Turkey varies significantly. The country can be divided into three major agricultural regions--the coastal areas, central Anatolia, and eastern Anatolia--with all three regions divided further into subregions to comprise the nine land use zones mentioned in Chapter 3. Production in the three major regions is given in Table 4.2.

The figures illustrate the relative importance of the coastal regions, which account for 50.0 percent of total agricultural production in Turkey. Central Anatolia accounts for 36.4 percent and eastern Anatolia for only 13.6 percent. While central Anatolia has equal percentages of plant and animal products, the share of plant products is higher in the coastal areas, and the share of animal products is higher in eastern Anatolia. While the production of grain, pulses, and animal products takes place all over the country, production of citrus fruits, olives, tobacco, cotton, sugar beets, and hazelnuts is mainly localized.

The coastal areas specialize in horticultural products and industrial crops. However, these areas also have a significant position in other farm products and in animal products. They account for more than one-third of the total wheat production and nearly one-third or more of other farm products. Central Anatolia specializes in the production of grains, livestock, sugar beets, and pulses, while fruits are almost nonexistent.

Eastern Anatolia, which is both generally and agriculturally backward, accounts for a relatively low share of all products, especially in industrial and horticultural products. The relative share of eastern Anatolia is highest, in livestock production, especially in cattle and goats, for which products it accounts for nearly one-fourth.

TABLE 4.2

Regional Distribution of Agricultural Production in Turkey, 1968 (percentages)

	Coastal Regions	Central Anatolia	Eastern Anatolia
Crops			
Wheat	34.3	54.2	11.5
Barley	29.2	55.5	15.3
Pulses	45.0	38.4	16.6
Tobacco	88.5	8.3	3.2
Sugar beets	31.9	63.2	4.9
Cotton	95.1	2.1	2.8
Hazelnuts	95.7	3.0	1.3
Grapes	52.8	34.1	13.1
Citrus	100.0	n.s.	n.s.
Olives	98.7	n.s.	1.3
Total	54.0	36.9	9.1
Livestock			
Sheep	29.2	42.4	18.4
Goats	48.8	26.8	24.4
Cattle	41.8	33.7	24.5
Milk	43.9	28.9	17.2
Meat	60.0	27.1	12.9
Total	44.0	35.2	20.8
Crops and livestock	50.0	36.4	13.6

Note: n.s. = Not significant amounts.

Source: Resat Aktan, "Basic Characteristics of Turkish Agriculture and Problems of Productivity" (Istanbul: Economic and Social Studies Conference Board, October 1972).

Because in Turkey land resources are relatively limited in relation to the agricultural population, and because there is the problem of underemployment and unemployment, both the labor-intensive products and the high-value products, such as fruits, vegetables, industrial crops, and intensive animal products, are likely to become more and more important, as indicated by the trends in the recent decades. The coastal areas, which have the greatest potential for irrigation development, seem to have the prospects for the most rapid development in high-value crops, and will most likely increase their share of total agricultural production value. Some writers have indicated that the coastal regions' share of products such as barley and pulses is likely to decline, since these products would be concentrated in other parts of the country. As mentioned previously, the percentage of irrigated land in the Marmara, Aegean, and Mediterranean regions that was used for wheat had a tendency to increase following the introduction of the Mexican wheat varieties. The share of total wheat production in these coastal regions has also increased, from 25 percent for the 1961-65 period to 32 percent for the 1969-71 period. This development has taken place mainly through a more rapid increase of yields than has occurred in central Anatolia and eastern Anatolia. In order to change this trend it would be of utmost importance to develop a higher yield in wheat technology in the two other regions.

PRODUCTION OF WHEAT AND OTHER MAJOR CROPS

The area devoted to wheat production in Turkey increased approximately 3 million hectares between 1948 and 1958. This increase occurred mainly in the steppe area, and in the early 1950s, as previously mentioned. The trend has since leveled off and, according to the agricultural census, around 9.0 million hectares have been devoted to wheat production in recent years, or about 55 percent of the annually sown area.

The area producing barley and rye has increased by about 1.0 million hectares during the last two decades. The area for barley production was about 2.8 million hectares in 1971 and for rye it was about 0.7 million hectares. The total area for small grains amounts to 12.5 million hectares; and most of the 8.5 million hectares of fallow land is rotated with these crops. These crops combined account for more than 80 percent of the cultivated area in Turkey.

The area devoted to maize has more or less remained constant during the two last decades at the level of 0.7 million hectares. The area for oats has also remained relatively constant, at the level of 0.4 million hectares. The rice area has doubled over the two last decades, but still represents only approximately 60,000 hectares. The total area in cereals comprises about 84 percent of the cultivated area in farm crops.

The area for industrial crops (sugar beets, tobacco, and cotton) and potatoes has doubled during the past two decades, and now occupies about 8 percent of the cultivated area for all crops. Much of the area devoted to industrial crops is irrigated. There has been an increasing use of irrigated land for growing cotton, to such an extent that it now dominates the cropping pattern of much of the irrigated area in the Mediterranean and Aegean regions. The development of the price relations between cotton and other crops competing for the land has also recently been in favor of cotton.

The area producing fruits and vegetables has nearly tripled during the last two decades and amounts now to about 1.5 million hectares. Approximately the same area is devoted to vineyards and olive groves. Added together with total crop area in Table 2.1 they give a total cultivated area of more than 27 million hectares.

The indexes of yield increase for four major crops are given in Table 4.3. The increase in yield for the last two decades for wheat and for barley is only about 40 percent, while for cotton and sugar beets yields more than doubled. The slow growth in yields of wheat and barley has been typical of all Near East countries. Rating the countries according to the growth of crop yields, from the 1961-65 period to 1971 Turkey ranks third in wheat and fourth in barley. For the world as a whole wheat yields increased only 44 percent from the 1948-52 period to the 1966-70 period. While the wheat yields in Turkey were 1.2 tons for 1966-70, the world average was 1.4 tons.

For cotton, which in some of the Turkish coastal areas competes with wheat, the story is different. Cotton yields have been increasing rapidly during the last decade and more rapidly in Turkey than in other countries. The average cotton yield in Turkey for the 1966-70 period was more than double the world average. Only a few countries have higher cotton yields than Turkey; in the Near East only Egypt has higher yields.

TABLE 4.3

Indexes of Yield Increases for Four Major Farm Crops in Turkey from 1946-50 to 1966-70

	Wheat	Barley	Cotton	Sugar Beets
1946-50	100	100	100	100
1951-55	123	131	95	130
1956-60	123	133	106	118
1961-65	125	129	162	146
1966-70	139	142	242	213
Annual compound growth rates, 1946-50 to 1966-70 (percentages)	2.0	2.3	4.6	3.4

Source: FAO, Production Yearbook (1946-70).

However, in some coastal areas wheat yields have been increasing more rapidly than cotton yields due to the introduction of the Mexican wheat varieties.

Regarding yields of sugar beets, the average yield in Turkey is a little above the world average, but not as high as in some Western European countries and in Israel.

The relatively rapid increase in Turkish yields of cotton and sugar beets is a result of technological advances, especially the improvement of varieties, the use of more fertilizer, and the improvement of cultural practices.

Crops normally grown on irrigated lands or in the higher-rainfall areas of the coast have also had increased yields. Dry beans, potatoes, onions, and melons are examples of such crops. Other crops, grown in lower-rainfall areas and on nonirrigated land using residual moisture, have had only slight yield increases. Examples of these are chick-peas, lentils, vetch, groundnuts, and some of the oilseed crops.

The yield of fruits has increased significantly in recent years, but because statistics are recorded on a tree-count basis it is difficult to determine how much of the increased yield has contributed to increased output.

The introduction of improved technology, including fertilizer, plant protection chemicals, and farm machinery, has lowered the level of crop yield fluctuations. A study by the State Planning Organization shows that the fluctuations have decreased for 10 of the 13 crops included in the study. In wheat the fluctuations had decreased from 9.5 percent to 6.2 percent and in absolute figures from 115 kilograms to 85 kilograms per hectare.

The production of wheat, the most important crop in Turkey, has increased significantly over a long period. Since 1950 wheat production has more than doubled, primarily because of an increase in cultivated area. Similar increases in the production of other small grains have occurred for the same reasons. Yearly production levels fluctuate in the rainfed areas, according to the amount and distribution of precipitation.

There have been only small increases in the production of maize, sorghum, and millets, which are produced mainly in irrigated areas where they have to compete with industrial crops and fruits and vegetables.

Cotton is one of Turkey's principal crops. In addition to supplying fiber for an expanding textile industry, cotton replaced tobacco in 1965 as Turkey's leading agricultural export. The cultivated area for cotton remained relatively stable for many years up to World War II, and a few years thereafter. Exports then were relatively small, and the slight upward trend in production was offset by rising domestic consumption. However, in the late 1940s and early 1950s cotton acreage rose dramatically, because of a strong demand both domestically and for exports, and because of favorable prices. During the last decade cotton acreage has been relatively stable at the level of 0.7 million hectares.

Cotton lint production increased from 58,000 tons in 1948 to 536,000 tons in 1971. This large increase can be attributed to the introduction of new high-yield varieties of upland cotton from Syria and the United States, to increased irrigation, and to the use of fertilizer and plant protection chemicals.

Practically all of the cotton in Turkey is grown in the Aegean and Mediterranean regions. The Cukurova area in Adana province is by far the largest cotton-growing area in Turkey, recording nearly two-thirds of the total acreage and half of the production. Half of the cotton in this area is grown under rainfed conditions. Cotton must compete with high-yielding wheats, citrus, vegetables, and sugar beets for the best irrigated land.

The Aegean area is the second most important cotton-producing area. The provinces of Aydin, Izmir, and Manisa provide more than three-fourths of the Aegean cotton acreage and production.

Cotton production in Turkey has helped to introduce a commercial concept into otherwise basic subsistence farming. Large-scale farming of cotton has contributed to the adoption of advanced cultural methods. Although machinery has for the most part replaced animal power in the preparation of land for planting, basin irrigation methods and poor land-leveling practices have inhibited a rapid shift to more complete mechanization. Through the improved land leveling and irrigation methods now being introduced yields might continue to increase.

The first successful sugar beet yield and sugar factory were established in Turkey in 1926. Three additional factories were added by 1934, and 14 more were built by 1965. It is reported that lack of fertilizer and irrigation water, along with the use of unimproved seeds, held down sugar beet yields during the 1950s (average yield was 18 tons per hectare).* However, since the mid-1960s substantial increases in yield have been recorded.

Sugar beets became a major industrial crop in the 1950s. After that production tripled, from 1.4 million tons in the 1952-56 period to 4.3 million tons in 1971, and the major reason for this jump was the significant increase in yields.

Apparently the cost of sugar production in Turkey in the early years was high. With increased yields the unit cost of production has decreased considerably, but sugar beets continue to be a labor-intensive crop. Further increases in production can be expected as a result of increased fertilizer use, better irrigation, and more high-yield varieties.

Mechanization of sugar beet production, except for land preparation, will probably be slow, because of the surplus labor available in the producing areas. It is important to note that increases in production have a significant impact on the output of by-product feeds for fattening livestock, such as sugar beet tops, pulp, and molasses.

---

*Edgar Z. Palmer, "Agriculture in Turkey: Long Term Projections of Supply and Demand," Istanbul (Roberts College: 1966).

TOBACCO AND OTHER MINOR CROPS

Tobacco has been grown in Turkey for over 360 years. The date of the development of the specialized variety called "oriental-type" tobacco is not known. Even though the oriental type is grown in Greece, Bulgaria, and Yugoslavia, Turkey is the leading producer of this tobacco, growing 35-45 percent of the total. Recently attempts have been made in Turkey to grow Virginia-type tobacco. Turkey now produces 3-4 percent of the world's tobacco, and its exports account for 7-10 percent of the world trade. Tobacco usually occupies less than 1 percent of the cultivated area of Turkey, but employs about 10 percent of the farm families; it is a highly labor-intensive crop.

Tobacco production increased from about 83,000 tons in 1948 to around 140,000 tons in 1971. The increase has been fairly steady, but some fluctuations have occurred. The increase has been due mainly to expanding area, not expanding yields. Yields have had a tendency to decline with the extension of tobacco cultivation to less suitable soils.

The best areas for quality tobacco are the Aegean and Black Sea regions, where about 55 percent and 27 percent of the nation's tobacco output are produced respectively. The average tobacco grown in the Marmara region is of poor quality and is largely used for blending.

Potato and onion production has doubled since the early 1950s. About half of this increase has resulted from higher yields and the other half from the increased area under production. Onions are produced in all regions, and though potato production is found in all regions, relatively more potatoes are grown in the Central North, Aegean, Black Sea, and Central South regions. Both these products are for the most part consumed locally.

Many vegetables are grown in home gardens in the villages, but most of the commercial vegetable production is centered around the cities, and particularly in the coastal areas. Commercial vegetable production has increased substantially in the past decade. Efforts are being made to develop an export market in Europe, particularly during the winter and early spring months. Future production can be expected to increase in the currently irrigated areas and in new irrigated areas being developed.

There has been only a limited increase in the production of dry legumes (dry beans, dry broad beans, chickpeas, lentils, and vetch). Production increased by about

100,000 tons in the recent 15-year period, or by about 25 percent. The area planted with these crops increased by about 10 percent during this period; thus yield increases have accounted for the remainder of the additional output. These crops are grown in the rougher, rainfed land and in small plots near the villages. Most of the output is consumed directly by the farmers or in nearby villages.

The increase in Turkey in the production of oil-yielding crops has been rather significant. The production of vegetable oils increased from 92,000 tons in 1948 to about 270,000 tons in 1968--a threefold increase. The largest increases in the production of these oils have been the result of substantial increases in the output of sunflower seeds, cottonseeds, and olives.

In addition to industrial uses vegetable oils furnish one of the most important elements in the local diet. The most important of the vegetable oils is olive oil, with the olive tree being native to Anatolia. The trend in the number of olive trees is sharply upward, and it is likely to continue upward, because the government has been promoting the expansion of olive orchards. Under the government programs wild olive trees are moved into the orchards, and grafted with improved varieties, with the farmer reaping the benefits. In most cases the land used is not suitable for other crops. Exports of vegetable oils have been small, and probably will continue small, while imports of some types of vegetable oils will increase.

A variety of fruits are produced in Turkey. The most important fruit crop is grapes. They are grown in all parts of the country, but are found mostly in the regions bordering on the Aegean and Mediterranean seas. According to state statistical data, grape production has more than doubled in the past twenty years.

Turkey is the world's largest commercial producer of dried figs, which account for substantial foreign earnings. Dried fig production has nearly doubled in the past two decades, from 25,000 tons in the 1948-52 period to 47,000 tons in 1970.

Most of Turkey's citrus fruit is grown in the Mediterranean coastal region under irrigation. According to official figures, the production of oranges and lemons has increased almost tenfold since 1950, and in recent years has been increasing at an average rate of about 10 percent per year. Further expansion is planned to capitalize on the growing European citrus market.

Production of deciduous fruits in Turkey has tripled since the early 1950s. Apple output has doubled in the past five years, and peach output has doubled over the past ten years. Continued expansion in the output of these fruits is expected.

The most important nuts produced in Turkey are hazelnuts, almonds, pistachios, walnuts, and chestnuts. A major nut-producing area is the Black Sea region, where about one-half of the output is grown. The number of nut trees increased from 140 million in 1950 to 219 million in 1969. The production of nuts has increased on the average about 8 percent per year during the past decade. Turkey is the world's largest producer of hazelnuts, and among Turkey's agricultural exports hazelnuts are third in value earned. The planted area for nuts, particularly for hazelnuts and pistachios, is gradually increasing, as are nut yields. With continued favorable prices production should continue to increase.

Though the production of tea is relatively unimportant in Turkey's total agricultural economy, it bears special mention because of the very rapid upward trends in tea production. Tea production was introduced in 1924, but no production of significance began until the early 1950s. During the 1952-56 period less than 1,000 tons were harvested, but by 1962 about 10,000 tons were produced and by 1971 the output was 34,000 tons. This production is all concentrated in the Black Sea region in an area squeezed between the high Pindar range and the sea. The area is almost subtropical with some 2,000 millimeters of rainfall per year, a good portion of which falls during the growing season. Though the tea industry is steadily expanding, it will continue to account for only a small portion of the total value of agricultural production.

Animal Production

Animals serve several purposes in Turkey. They are used for motive power, and provide significant quantities of milk products, meat, wool, mohair, hides, and skins. The growing demand and the deep-seated traditions of the Turkish people and the nature of the land in many of the provinces are such that animal production is destined to remain an important part of the agricultural economy. However, the livestock sector has received relatively little government support. Livestock fulfill a role largely as residual users of land, especially of the natural grazing area of approximately 19.3 million hectares, which is 24.5 percent of the total land area.

Most of Turkey's livestock are handled in the traditional village manner. Communal pasturing is carried out on noncropped land around the villages. In addition the livestock graze on the residues of each year's crop. At present very little supplemental feeding of forage is carried out, and virtually no feeding of grain. Yet two forces have been at work: on one hand the feed base has been steadily decreasing, and on the other hand the numbers of animals have been increasing. Moreover the income elasticity created by the demand for meat is strongly positive, and a more populous and prosperous nation will demand more and more meat. In recent years, however, both the rate of decrease in feed resources and the rate of increase in the number of animals have slowed down. Also the need for animals for work purposes is slowly decreasing. Thus the need to feed horses, donkeys, mules, and camels will decrease, making more feed available for meat- and milk-producing animals.

Unfortunately this buildup of overall animal numbers, along with the reduction in land area committed to the production of animal feed and available grazing, has created circumstances wherein the livestock activity is neither highly productive nor highly efficient. Available feed supply per animal unit appears to be actually below the maintenance requirements for the animal population during much of the year. Product output per animal unit is quite low as compared to the product output per animal unit of many of the important animal-producing countries. The average yield of cows' milk in Turkey is only around 600 kilograms per year, and has increased very little in the last two decades.

A number of serious problems face the livestock industry in Turkey. With the number of animals increasing, the feed supply, particularly forage and pasture, in most parts of the country has been decreasing, thus creating serious nutritional problems in livestock production. There is little reason for optimism concerning the livestock-feed relationship in the next few years. It is unlikely that any substantial changes in range and forage production can be made in the near future, as Turkish grazing tradition inhibits the possible voluntary reduction in animal numbers.

Another important problem is the predominance of indigenous breeds that have low productive and reproductive capabilities, and thus have low input-output efficiencies for producing meat, milk, and other livestock products. Selection and breeding programs are needed to combat this problem. Other possible solutions are the introduction of selected breeds from outside Turkey for crossing with indigenous breeds, and the raising of pure-bred herds. This is particularly needed for dairy and poultry production.

It has been estimated that the animal losses in Turkey due to disease and to parasites amount to about TL 2.0 billion per year. This constitutes about a 20 percent loss in potential income from the livestock industry. A vigorous and accelerated animal health program is needed to reduce this loss.

A large portion of the grazing land is made up of the communal pastures used by village herds and flocks. Under the existing system villagers send all the livestock they own to the communal pastures. The villagers who own the largest numbers of animals derive a proportionately greater benefit from the communal pastures, thus encouraging all villagers to increase their livestock numbers, and in turn causing a further depletion of the already scarce forage resources.

Another potentially important means of alleviating the growing shortage of animal protein in Turkey is by developing the poultry industry. The technology of modern poultry production in Turkey for both eggs and meat is well developed and highly transferable; geography and space are not limiting factors; technical know-how, feed, and capital can be imported; labor is relatively cheap in Turkey; and present prices for poultry meat and eggs are very favorable.

The relative shortage of animal products in Turkey has led to an increase in their prices, especially in meat. Retail meat prices in Istanbul and Ankara almost doubled in the six-year period from 1962 to 1967.

Forest Production

Although in the past there were forests over most of Anatolia, a journey today over the interior of Turkey might give one the impression of a land without trees. Yet there are forests in certain interior regions, and on the whole Turkey has substantial forest resources. Approximately 18.0 million hectares, or 23.7 percent of the total land area, is classified as forest area, of which between 50 and 60 percent is high forest. The forest resources have been greatly underestimated and underexploited.*

---

*Edward J. Cohn, "Turkish Forest Products: Problems and Opportunities" (1967).

The current estimate of the annual growth of forests without bark is 12.6 million cubic meters (excluding fuel-wood), which is removed at a present rate of about 8.3 million cubic meters. This compares with a takeoff of 4.3 million cubic meters in 1968, or less than one-third of the annual increment.

The development of forestry and forest industries could make a significant contribution to Turkey's foreign exchange earnings if export could take place at competitive prices. Expansion of the Turkish road network is an important factor in forestry development, and large investments are needed to establish an industrial complex. A revision will have to come in the timber price policy, since the current prices are far above those of the major timber-producing countries. Other needs include better equipment for timber extraction and for reduction of waste, the development of logging into a year-round operation, and the improvement of the quality of timber products.

Efforts to shape the Turkish forestry policy should not ignore the farmers in the so-called forest villages. Some seven million people, or about one-third of the agricultural population live in or near villages that are, by legal definition, forest areas. These forest villages are mainly poor mountain villages with very limited areas of agricultural land. Their average per capita income is estimated at only one-sixth of the per capita income for the country. Much of the meager income in these villages is derived from goats.

Agricultural Exports and Imports

Turkey exports many agricultural crops, and in recent years approximately 80 percent of her total exports has consisted of agricultural products.

The largest group of Turkish agricultural exports are the industrial crops (cotton and tobacco) with dried fruits (figs and raisins) next in order. As a result of the increase in cotton production, cotton by-products--cottonseed cake and meal--have also become relatively important exports.

The value of total Turkish exports of food and other agricultural products has been increasing at an average annual rate of about 5 percent per year. The proportion of cotton exported has varied between 50 and 60 percent of the annual crop. The amount of tobacco exported has nearly doubled since 1948, with the quantity exported representing about 50 percent of the tobacco produced in Turkey each year. Ranking as the world's leading producer of hazelnuts, as mentioned above, Turkey is also the leading exporter of hazelnuts.

The Turkish export of dried figs is also important, but the quantity exported has only increased slightly. Figs now rank fourth in Turkey in terms of value of exports. Significant increases in the export of citrus fruits have occurred during the past 20 years, and further increases are expected as production increases in the newly irrigated areas of the southern coast.

An inadequate export performance is one of the serious obstacles to the future development of the Turkish economy. Unless exports are increased the country will be unable to finance the volume of imports required for her current consumer needs and investments and, at the same time, repay foreign debts.

The growth of total Turkish imports has been rapidly increasing since the late 1950s. However, the portion of total imports that are classified as food and agricultural products has sharply declined from 11 percent in 1954 to less than 4 percent in 1968.

Agricultural Income

The total agricultural income in Turkey (including forestry and fishing) in 1970 was TL 38.3 billion, according to current prices (TL 24.3 billion, according to 1961 prices). This represented 30.8 percent of the Turkish national income in 1970. But it means a sharp decline from 1927, when agriculture held a 67.0 percent share of the national income. The decline has been especially rapid during the last decade.

The agricultural income per capita (agricultural population) is approximately half the average per capita income in Turkey and less than one-fourth of the per capita income in the nonrural sector.

A study of Turkish agricultural incomes by Eva Hirsch covering the years 1951 to 1953 indicates that the average annual income among the 10 percent of farmers earning the lowest income was TL 278, while the 10 percent with the highest income earned an average of TL 8,926.* The study

---

*Eva Hirsch, Poverty and Plenty on the Turkish Farm (New York: 1970).

found the agricultural average for Turkey as a whole to be TL 4,367. The 10 percent in the lowest income group had only a 1.0 percent share of the total agricultural income, and the highest 10 percent had a share of 52.8 percent. The 80 percent of farmers earning low incomes had a one-third share of agricultural earnings, and the remaining 20 percent with high incomes had a two-thirds share.

Data from a study by the Turkish State Planning Organization, covering Turkish agricultural incomes for 1963 and published in the Second Five-Year Development Plan, show that small enterprises, which accounted for 68.8 percent of all agricultural enterprises, had a 24.8 percent share of the agricultural earnings, with an average income of TL 2,900 per enterprise and TL 485 per agricultural worker. The medium-sized enterprises, which accounted for 27.5 percent of all enterprises, had 42.0 percent of the agricultural earnings, with an average income of TL 10,300 per enterprise and TL 1,117 per worker. The large enterprises comprised 3.6 percent of the total; their share of total earnings was 23.2 percent, their earnings per enterprise were TL 44,510, and their earnings per worker were TL 7,417. The largest enterprises represented 0.1 percent of the total, and their income share was 10.0 percent of the total. Their average earnings were TL 298,500 per enterprise and TL 49,750 per worker.

A recent study for 1968 confirms the findings of the Hirsch study. It shows that 59.1 percent of all farm families, with an average income of less than TL 5,000 per year, accounted for 18.8 percent of total agricultural earnings; 23.5 percent of farm families, at the income level of TL 5,000-10,000, had 19.9 percent of the total; and 0.7 percent of the families, earning above TL 100,000, had 8.3 percent.* According to the same study, 10 percent of the farm families with the lowest income earned 1.0 percent of the total agricultural income, and the 10 percent with the highest income earned 49.0 percent.

All three studies confirm the following facts:

1. The income of the majority of Turkish farm families appears to be below a reasonable poverty level.

---

*Tunker Bulatayit et al., "Income Distribution in Turkey, 1968," Faculty of Political Sciences, Ankara University: 1971; (in Turkish).

2. There is a marked income disparity between the agricultural and industrial sectors.
3. There are marked inequalities in income distribution within the agricultural sector.
4. There are great fluctuations in farmers' income from year to year.
5. In areas at particularly high altitudes, where the opportunities within agriculture are severely limited and the population dependent upon it is increasing, there is the danger of a long-term decline in real living standards, unless the dependent population is reduced.

CHAPTER 5
THE PRODUCTION INPUTS

SEED

The state farms in Turkey have played a significant role in the country's grain seed programs since 1943. A law controlling the certification of seeds was passed in 1963. Subsequently a Directorate of Regional Variety Testing was set up under the Ministry of Agriculture, and was given the task of conducting seed experiments and certifying new varieties. Included among the grain seeds produced and distributed in Turkey are the following classes: elite seeds, foundation seeds, registered seeds, and certified seeds. Elite and foundation seeds are produced by the state research stations. Registered seeds are grown mainly by the 21 state farms. Certified seeds are produced both by the state farms and by contracting farmers. The contracting farmers work under the supervision of the state farms and are responsible to them.

The certification of seeds is carried out by seed testing stations, located in Ankara, Istanbul, and Izmir. These stations provide trained field inspectors, and their laboratories are equipped for testing seeds for purity and germination according to the standards laid down in the seed certification law. Thus the essential facilities for an effective seed program are available. The state farms are well distributed in all parts of the country and are well equipped with field machinery. Overall they have the capacity to produce about 120,000 tons of certified wheat seed each year.

According to the present production pattern the country's need for seed has been estimated at approximately 2 million tons per year, of which 1.4 million tons are wheat seeds. Since seeds of cereals would have to be renewed only every four to five years, the annual requirement for improved seed and controlled seed of various types has been estimated at approximately 0.4 million tons.

The General Directorate of the Ministry of Agriculture has the responsibility of distributing three of the classes of seeds--foundation, registered, and certified--and of preparing annual improvement programs for seeds of various crops. The seed varieties to be produced, the regional distribution of seeds, and prices are determined by the National Committee for Improved Seed.

Table 5.1 gives the amounts of certified wheat seeds produced and distributed to Turkish farmers in recent years.

TABLE 5.1

Production and Distribution of Certified Wheat Seeds in Turkey, 1965-72
(tons)

Year	Production	Distribution
1965	97,038	76,305
1966	106,927	127,310
1967	163,607	210,091
1968	257,645	162,874
1969	253,308	226,988
1970	325,000	94,025
1971	226,919	38,707
1972	175,000	91,955*

*Provisional

Note: The differences between production and distribution are due to carry-over or to seed supplied by Soil Products Office (TMO).

Source: Ministry of Agriculture.

The seed improvement programs have been very useful in distributing to farmers those seed varieties which are best adapted to local conditions, and which have facilitated the introduction of high-yield wheat varieties, initiated in 1967.

There are various reasons why the targets that have been set for distributing certified wheat seed have not always been achieved: The production of certified seeds in some years has been lower than planned; the targets set have been overoptimistic; the organization of the distribution activities has not been efficient; the availability of credit to the supplying organizations and to the farmers has not been sufficient; and especially after bad crop years many farmers have been in debt to the credit organization and consequently could not get new loans.

Prices of certified wheat seeds are generally fixed at a level that is 20-30 percent higher than the regular wheat prices. The government also subsidizes part of the seed transportation costs. Still the farmers complain that the prices for seeds are too high. However, the inadequacy of credit and the freezing of credit for farmers who are in debt seem to have been the major bottlenecks.

FERTILIZER

The use of large quantities of fertilizer represents a relatively recent development in Turkey. In the late 1950s the use of fertilizer was pioneered by Turkish scientists and institutions. Subsequently the government, recognizing the value of fertilizer, authorized the construction of two superphosphate plants and one ammonium sulphate plant.

The consumption of fertilizer increased rather slowly during the 1950s and the early 1960s. To step up consumption and thereby help meet the food requirements of a fast-growing population, the government took several measures during the period of the First Five-Year Plan (1963-67). These measures resulted in rapid increases in fertilizer consumption; they can be summarized as follows:

1. The government improved research and extension services, aiming to facilitate the use of fertilizers by farmers. For this purpose a plant nutrition department was formed in the Ministry of Agriculture. A special working group on fertilizer was formed to train technical personnel and farmers. All fertilizer specialists and regional agents were trained through short courses. An intensive farmers' training program on soil fertility and fertilizer use was also carried out.

2. With the assistance of FAO approximately 15,000 fertilizer demonstrations, covering the major crops, were given throughout the country.

3. A special fertilizer fund was established in the Agricultural Bank of Turkey. This fund has been used to supply fertilizer credit to the farmer and to finance the supplying agencies.

4. The retail prices of the most important types of fertilizer were set at the same level for the whole country, and were gradually lowered.

5. Government agencies improved fertilizer distribution. Until 1967 the Agricultural Supply Organization (T. Z. Donatim) was the only government distributor for straight phosphorus and nitrogenous fertilizers. In 1967 local credit cooperatives, which are coordinated and controlled by the Agricultural Bank of Turkey, and other large cooperatives were authorized to distribute fertilizers. Private companies were allowed to handle the potassium and compound fertilizers. (The estimated retail sales of fertilizer through various agencies for 1971 are given in Table 5.2.)

TABLE 5.2

Estimated Percentages of Turkish Retail Sales of Fertilizer by Various Agencies, 1971

Agency	Distributing Company	Percent of Sales
Ministry of Commerce	Sugar Bank, through sugar beet cooperatives	5
	Agricultural Bank, through credit cooperatives	23
	Production and sales cooperatives	9
Ministry of Industry	Azot Sanayii Factory, direct sales	4
Ministry of Agriculture	T. Z. Donatim	51
Private sector	Dealers and importers*	8

*This group was denied import permits for 1971, but in 1970 it distributed 16 percent of total sales of fertilizer.

Source: John M. Hill, "The Need for Reorganization of Turkey's Fertilizer Industry" (AID: Ankara, February 2, 1971).

Currently fertilizer is largely imported by Turkey and distributed through government agencies, of which the state-owned Agricultural Supply Organization, T. Z. Donatim, is the most important. This agency receives instructions from the Ministry of Agriculture and has warehouses located throughout the provinces. In addition the Agricultural Bank imports fertilizer directly and resells it through the local credit cooperatives, about 2,000 of which are scattered throughout Turkey. A number of production and marketing cooperatives also distribute fertilizer to their members.

The sharp increase in fertilizer consumption in Turkey during the last two decades can be seen from the data presented in Table 5.3. The data show that consumption of fertilizer has increased more than 40 times compared with the early 1950s. From 1961 to 1971 the annual compound rate of increase was 29 percent.

TABLE 5.3

Consumption of Commercial Fertilizers in Turkey, 1950-72
(tons of nutrients)

Year	Nitrogen	Phosphorus	Potassium	Total
1950-54	5,722	4,708	3,158	13,567
1955-59	12,365	7,234	1,304	20,903
1960	9,217	9,582	652	19,451
1961	28,000	12,320	14	40,334
1962	37,352	16,724	5,029	57,816
1963	37,352	34,986	10,080	82,418
1964	51,000	42,400	4,320	98,320
1965	68,000	70,400	3,120	141,520
1966	95,900	90,700	3,600	190,200
1967	138,100	132,200	8,600	278,900
1968	186,600	181,700	11,700	380,000
1969	231,100	200,600	14,300	446,000
1970	243,000	175,900	12,700	431,600
1971	286,647	194,694	13,247	494,588
1972	354,353	245,999	27,173	627,525

Note: The figures for the 1950-54 period and the 1955-59 period are averages.

Source: FAO, Production Yearbook (1950-1972) and Statistical Yearbook of Turkey, 1973, for figures for 1972.

According to the Second Five-Year Plan, only 5 percent of the acreage of wheat, rice, sugar beets, citrus, and tea was estimated to be fertilized in 1967. Overall only 18 percent of the cultivated land received nitrogen and 24 percent received phosphates. The average amount of nutrients received per hectare was only 7.2 kg. However, there has been a dramatic increase since 1963 in the percentage of wheat land. The amount of fertilizer used for wheat in Turkey has increased dramatically since 1963. The percentage of the total wheat area fertilized with nearly the recommended rates has increased from 3 to 29 percent. This is shown in Table 5.4, which also indicates that the tons of nutrients consumed by wheat have increased more than 12 times at a compound annual growth rate of about 37 percent. Between 40 and 50 percent of the total fertilizer consumption in Turkey is now recorded by wheat. While the percentage of nitrogen fertilizer used in wheat was only 14 percent in 1963, it has recently increased to more than 40 percent.

In 1968 Turkey was producing only about one-fourth of its total fertilizer requirements. This situation created problems in the financing of expenditures in foreign currency, in the timing of fertilizer imports, and in distribution and usage. In order to meet increasing demands the government has given great emphasis to domestic fertilizer production, and in the future supplies will be obtained mostly from national sources. An investment of about $120 million was made in fertilizer production facilities during the period from 1968 to 1970. This resulted in the construction of three new plants, the expansion of another plant, and the conversion of a plant. With these additional facilities, the production of nitrogen fertilizer is expected to increase ten times over the figure for 1968, and phosphorus fertilizer production is expected to increase four times.

During the same period a major switch from the use of low-analysis fertilizer to high-analysis fertilizer took place. Triple superphosphate was produced and used instead of single superphosphate. In addition diammonium phosphate fertilizer was introduced for wheat and other cereals, and two plants were constructed for the production of diammonium phosphate. Urea has also been introduced as a fertilizer, and the government is planning to build a 1,000-ton-a-day urea-producing plant.

Obviously the old fertilizer factories in Turkey were too small to produce at low costs. They also produced expensive types of low-analysis fertilizer. Thus

TABLE 5.4

Fertilizer Consumption for Wheat in Turkey, 1963-71

Year	Nutrients (thousand tons)		Percentage of Total Wheat Area Fertilized with Nearly Recommended Rates	Percentage of Total Fertilizer Consumption	
	Nitrogen	Phosphorus		Nitrogen	Phosphorus
1963	5.3	12.7	3.0	14.2	36.3
1964	6.3	15.3	3.8	12.2	36.0
1965	10.5	29.1	7.3	15.4	41.3
1966	16.8	35.0	8.5	17.5	38.6
1967	42.8	74.5	17.5	31.0	56.4
1968	64.9	113.9	24.2	34.8	62.7
1969	99.5	114.9	25.1	43.1	57.3
1970	101.2	84.0	24.6	41.6	47.8
1971	118.6	102.1	29.3	41.3	52.4

<u>Source</u>: Ahmet Demirligakmak, director of Wheat Research and Training Center (Ankara).

the cost of fertilizer produced by domestic companies has been very high, and has been subsidized by the government to provide a balance with the prices of imported fertilizer. The devaluation of the Turkish lira in 1970 led to a further increase of 16 percent in controlled fertilizer prices. (See Table 5.5.)

TABLE 5.5

Turkish Retail Prices of Fertilizers, 1960-72
(kurus per kilogram)

Year	Ammonium Sulfate, 21 percent	Ammonium Nitrate, 21 percent	Superphosphate, 19 percent	Potassium Sulfate, 50 percent
1960	55	n.s.	50	85
1961	50	n.s.	50	85
1962	47	60	44	85
1963	63	60	44	75
1964	63	60	44	80
1965	63	60	45	100
1966	59	56	40	100
1967	59	56	37	90
1968	54	54	37	80
1969	54	54	37	90
1970	60	60	45	100
1971	60	60	45	120
1972	60	60	45	160

Notes: Prices of fertilizers (with the exception of potassium sulfate) are determined by the order of the government, and are the prices for the entire country. n.s. = Not significant amounts.

Source: Ministry of Agriculture.

Since Turkey is short of suitable raw materials, including natural gas, for ammonia production, only one plant is carrying out ammonia synthesis. The other plants will require imported ammonia. Due to the low content of Turkish rock phosphate, this raw material will also have to be imported, either in the form of rock phosphate or phosporic acid. As yet only small quantities of potassium are needed, and the total amount is imported.

Only fertilizer supplied by the Agricultural Supply Organization and by the agricultural credit cooperatives is financed and distributed to the farmers as credit in kind by the Agricultural Bank, and the bank is the only institutional source of credit for fertilizer. The recent credit situation for fertilizer distributed to the farmers is indicated by Table 5.6.

Fertilizer research is carried out by various government ministries, through their agencies, as indicated in Table 5.7. Budgetary problems have limited research activities, and previous research has concentrated on the use of low-analysis fertilizer at relatively low rates of application, and on the traditional varieties of low-yield grains.

The Sugar Company appears to have an extensive fertilizer research program on sugar beets. The fertilizer research by universities is carried out by the agriculture faculties of the University of Ankara, the University of Izmir, and Attaturk University, but their activities have been limited by restricted budgets.

Extension services, including recommendations on the use of fertilizer, are also provided by agencies of the various ministries, as well as by private groups, as indicated in Table 5.8.

Farms in the southern and western coastal regions, where the best agricultural lands are under irrigation, consume large quantities of both nitrogen and phosphate fertilizers. The Black Sea area consumes relatively large quantities of nitrogen, but only a small amount of phosphate. On the Anatolian plateau phosphate is currently being used on a wide scale, while the use of nitrogen for wheat, which is the area's main crop, has not yet been widely established.

In the coastal areas highly favorable cost-benefit ratios for fertilizer used on vegetables, fruit crops, cotton, and Mexican varieties of wheats make fertilizer a good investment for farmers. In these areas fertilizer credit repayments by farmers are running at the relatively high rate of 75 percent to 98 percent, while central and southeastern Turkey show large amounts of overdue debts. Production and sales cooperatives play a major role in fertilizer sales in the coastal areas.

In spite of the rapid increase in fertilizer consumption since 1960, the rate of application is still low. The prospects for increasing the rate of application are particularly strong in the irrigated and high-rainfall areas, but this growth will require a continuous stream of new high-yield varieties of crops that respond well to fertilizer consumption.

TABLE 5.6

Distribution of Fertilizer Loans by Agricultural Bank of Turkey, as of December 31, 1969

Method of Distribution	Interest (percent)	Amount of Loans (thousands of TL)	Percentage of Loans
Directly by Agricultural Bank	9	407,908	52
Through credit cooperatives of Agricultural Bank	7	351,112	45
Through sales and production cooperatives	8	23,721	3
Total		782,741	100

Source: John M. Hill, "The Need for Reorganization of Turkey's Fertilizer Industry" (AID: Ankara, February 2, 1971).

TABLE 5.7

Fertilizer Research and Soil Testing in Turkey

Ministry	Agency	Number of Stations or Laboratories Working on Fertilizer Research	Number of Soil Fertility Testing Laboratories
Village Affairs	General Directorate of Soil and Water Conservation	5 (plus 1 soil and fertilizer research institute	24
Education	Universities	3	2
Agriculture	n.r.	8	0
Forestry	n.r.	0	1
Commerce	Sugar Company	2	1

Note: n.r. = Heading not relevant.

Source: John M. Hill, "The Need for Reorganization of Turkey's Fertilizer Industry" (AID: Ankara, February 2, 1971).

TABLE 5.8

Fertilizer Extension Services for Turkish Farmers

Government or Private Services	Agency
Ministry of Commerce	Sugar Company
Ministry of Village Affairs	General Directorate of Soil and Water Conservation
Ministry of Agriculture	Technical Agricultural Organization
Private sector	Importers and dealers (recommendations for use of compound fertilizer)
Cooperative sector	Production and sales cooperatives (recommendations for use of fertilizer on specific crops, such as tea, hazelnuts, etc.).

Source: John M. Hill, "The Need for Reorganization of Turkey's Fertilizer Industry" (AID: Ankara, February 2, 1971).

Together with an increase in the irrigated area from about 6 percent to about 33 percent of the total cultivated area (1.5 million hectares to 8.0 million hectares) and improved water management, the rapidly increasing use of fertilizers represents one of the main factors in Turkish agricultural development. Together these factors may have a tremendous long-term impact on production.

As indicated in Table 5.9, which gives data on fertilizer prices and on fertilizer-wheat cost ratios, it takes on the average slightly more than 3 kg of wheat to pay for one kg of plant nutrient.

The most favorable price ratios occur in the high-analysis fertilizer and in the compound types of fertilizer. How the price ratios in Turkey compare with the ratios in other countries is shown in Table 5.10.

Regarding the benefit-cost ratios for wheat by use of fertilizer, the ratio, when the weather is favorable, is approximately 4:1 for the high-yield Mexican wheat varieties in the coastal areas, and 3:1 for the high-yield imported winter wheat varieties. By comparison the ratio is 2:1 or less for the native varieties. Ratios for the Mexican and native varieties are given in Table 5.11.

TABLE 5.9

1971 Fertilizer Prices and Plant Nutrient-to-Wheat Cost Ratios in Turkey

Type of Fertilizer	Product Cost (TL per ton) Factory	Retail	Price of Nutrient (TL per ton)	Cost of 1 kg of Nutrient in kgs of High-Yield Wheat (at 90 kurus per kilogram)
Ammonium sulfate, 21 percent nitrogen (paper bags)	480	600	2,857	3.17
Ammonium sulfate, 21 percent nitrogen (jute bags)	520	640	3,047	3.39
Ammonium nitrate, 20.5 percent nitrogen (paper bags)	480	600	2,927	3.25
Ammonium nitrate, 26 percent nitrogen (paper bags	580	710	2,731	3.03
and plastic bags)	620	750	2,884	3.20
Superphosphate, 17 percent phosphate (water-soluble in paper bags)	400	450	2,647	2.94
Triple superphosphate, 43 percent phosphate (water-soluble in paper bags	930	1,000	2,326	2.58
and plastic bags)	970	1,040	2,487	2.76
Diammonium phosphate (18-46-0 in plastic bags)	1,200	1,350- 1,500	2,031- 2,343	2.26- 2.60
Urea, 46 percent nitrogen (plastic bags)	imported	1,200	2,609	2.90
Compound, 20-20-0 (plastic bags)	imported	1,100	2,750	3.06

Note: Prices for ammonium sulfates, ammonium nitrates, superphosphate, and triple superphosphate are controlled by the government and apply to the entire country. Prices for other fertilizers are not controlled.

Sources: Ministry of Agriculture, Turkey; and John M. Hill, "The Need for Reorganization of Turkey's Fertilizer Industry" (AID: Ankara, February 2, 1971).

TABLE 5.10

Kilograms of Wheat Required to Purchase One Kilogram of Fertilizer Nutrient in Various Countries at Prices Prevailing in 1966/67

Country	Kilograms of Nitrogen (N)	Kilograms of Phosphate (P_2O_5)
Greece	2.1	1.5
Taiwan	4.5	2.5
Mexico	3.4	2.2
Colombia	3.4	3.5
Brazil	5.3	2.4
India	4.1	3.4
Turkey[a]	3.3	2.7
Iran[b]	3.6 (2.6)	3.6 (2.6)

[a]Figures for Turkey compiled by John Hill (of AID) using September 27, 1970 prices for fertilizer and high-yield wheat.

[b]Figures for Iran compiled by the author from paper by Hill and Rhodes, "Fertilizer and Iranian Agriculture," based on a wheat price of 6,000 rials per ton and on unsubsidized fertilizer. Based on 20 percent fertilizer subsidy, only 2.6 kg of wheat would be required to buy one kg of nutrient. (75 rials = $1 US).

Source: U.S., Department of Agriculture, Economic Research Service, Economic Progress of Agriculture in Developing Nations, 1950-68.

The recent shortfalls in fertilizer in Turkey, compared with consumption targets set by the government, have mainly been due to financial difficulties--both with Turkish lira and foreign exchange--in covering manufacturing costs, imports, wholesale procurement, distribution costs, and retail credit.

Inefficient coordination of fertilizer programs also has hurt the Turkish fertilizer industry. With four government ministries--Industry, Commerce, Agriculture, and Village Affairs--handling the major programs, the industry has become fragmented into four independent sectors: production, credit, distribution, and research and extension. Each segment pursues its own function and is compensated independently. This fragmentation, rather

TABLE 5.11

Benefit-Cost Ratios for Wheat in Turkey Based on
FAO Fertilizer Demonstrations, 1963-65

Variety of Wheat	Year	Number of Farms	Phosphate	Nitrogen and Phosphate	Nitrogen Phosphate and Potash
Mexican	1966	106	n.a.	4.0	n.a.
Native dryland	1963	979	1.8	1.7	1.7
	1964	1,924	2.2	1.7	1.2
	1965	2,365	2.0	2.4	1.8
	1963-65 average	5,268	2.0	1.9	1.6
Native irrigated land	1965	42	n.a.	1.8	2.0

Note: n.a. = Data not available.

Source: Report of the CENTO Ad Hoc Working Party on Fertilizer, Islamabad, (Pakistan, September 1968).

than benefiting Turkish farmers and consumers, has tended to serve the interests of each of the four groups. It has led in part to fertilizer shortages, credit problems at various levels, and less-than-optimum fertilizer recommendations to farmers.

The limited margin for the distribution of fertilizer represents still another problem for the industry. The margin between the manufacturer's fixed selling price and the fixed retail price is not adequate to cover average distribution costs. In particular long-distance transportation to central and southeastern Turkey can result in large losses on all fertilizer products with the possible exception of diammonium phosphate. Under the present system the losses will be absorbed by the Agricultural Supply Organization, since production and sales cooperatives and private dealers are not able to effectively and economically supply fertilizer and extension and credit services to farmers in the central and southeastern areas.

CHEMICALS FOR PLANT PROTECTION

The different types of chemicals used for plant protection in Turkey include pesticides, chemicals for disease control, weedicides, and chemicals to prevent the premature falling of fruit. Until the early 1960s the government, through the Ministry of Agriculture, carried out most of the pest and disease control activities, and the crops that received protection were mainly the industrial crops, fruits, grapes, and olives.

Under the First Five-Year Plan (1963-67), the government continued to play a primary role in pest and disease control, but in addition the farmers were encouraged to develop their own plant protection program, under the government's control. Actually the government's plant spraying program, administered by the General Directorate of Plant Protection, never reached most of the farmers, and the program has now been largely abandoned.

Under the Second Five-Year Plan (1968-72), the government conducts pest control activities and fights the diseases that are likely to spread rapidly and create large-scale plant losses. Beyond this the state provides leadership in organizing the farmers' efforts, conducts inspections, and provides technical aid and equipment. In addition the government continues to distribute pest control chemicals.

The growth of plant protection work in Turkey after 1964 is shown by the number of sprayers and dusters used over the last decade (see Table 5.12). This number virtually doubled from 1963 to 1967. However, plant protection efforts seem relatively limited when one realizes that there are less than 250,000 sprayers and dusters available for about 3 million farm units.

TABLE 5.12

Number of Sprayers and Dusters in Turkish Agriculture, 1961-72

Year	Sprayers	Dusters
1961	29,400	9,700
1962	32,800	12,500
1963	64,800	21,200
1964	96,400	22,900
1965	104,300	24,800
1966	133,900	27,600
1967	134,200	34,600
1968	153,800	36,400
1969	154,600	47,800
1970	151,835	39,332
1971	189,070	44,053
1972	182,602	46,699

Source: State Institute of Statistics

FARM MECHANIZATION

A large proportion of the farms in Turkey are small and fragmented, often containing irregularly shaped fields and stony patches: the primary sources of motive power are still bullocks and horses. Compared with most developed countries the degree of mechanication in Turkey is low and there continues to be a heavy reliance on human labor. Many of the small farms have only a wooden plow and a wooden drag pulled by draft animals. There are still about 2 million wooden plows in use in Turkey, indicating that about two-thirds of the number of operational farm units are totally, or at least partly, relying on such plows. Spraying equipment is usually available only for special crops. Most of the grain is threshed by having animals tread on the threshing floor.

According to the data in Table 5.13, the number of tractors in Turkey increased from 2,000 in 1948 to 40,000 in 1955, and to 116,000 in 1971. The area cultivated by tractors in 1971 amounted to an estimated 8.7 million hectares, or more than one-third of the total area under cultivation. However, the high rate of tractor mechanization, especially from 1948 to 1955, did not result in a reduction in the number of draft animals. In that period there was also a large increase in total area cultivated--7 million hectares, an increase of 50 percent. The area cultivated by animals increased approximately 4 million hectares, and the area cultivated by tractors increased approximately 3 million hectares. Tractors and other types of farm machinery, in addition to spurring the rapid extension of cultivated area, facilitated planting and harvesting at optimum periods. However, as previously mentioned, mechanization also led to the plowing up of marginal and even submarginal land that should have been kept as grazing land. As soon as the period from 1948 to 1955 of exceptionally good weather ended, the mechanization boom ceased, and the number of tractors became relatively fixed.

After 1963 the rate of mechanization picked up again, partly because of the new impetus given to agriculture by the five-year plans. The number of tractors increased 100 percent during the 1962-69 period; however, the trend in the number of draft animals did not start to show a decline until the end of the period. On the other hand the area cultivated by draft animals has been declining since 1964.

There are very few two-wheel tractors in Turkey. The most common are the four-wheel tractors with a 35-to-50 horsepower engine. Thus the impact of tractor mechanization on Turkish agricultural development appears to have been less than optimal because of the way it has taken place. An ideal strategy would have called for new technology to be introduced as part of a well-tested package. It was unfortunate that the first 40,000-50,000 tractors seem to have been introduced in Turkey without much understanding of this principle. The equipment for soil cultivation that was imported along with the tractors, including moldboard plows, disk plows, and disk harrows, had at that time already been recognized in the United States as being obsolete and unsuitable for dry-land farming. In the United States the chisel plow, which does not turn the soil, had been almost universally adopted. Actually in many areas of Turkey the wooden plow would lessen

TABLE 5.13

Number of Tractors and Draft Animals in Turkey, and Area Cultivated, Selected Years

Year	Number of Tractors (thousand)	Area Cultivated by Tractors (million hectares)	Pairs of Draft Animals (million)	Area Cultivated by Draft Animals (million hectares)	Total Cultivated Area (million hectares)	Percentage of Total Crop Area Cultivated by Tractors
1948	2	0.1	2.4	13.8	13.9	1
1950	17	1.2	2.5	13.3	14.5	9
1955	40	3.0	2.6	18.0	21.0	14
1957	44	3.3	2.6	18.9	22.2	15
1960	42	3.2	2.6	20.1	23.3	14
1961	43	3.2	2.6	19.9	23.1	14
1962	44	3.3	2.7	20.0	23.2	14
1963	51	3.8	2.7	20.0	23.9	16
1964	52	3.9	2.7	20.0	23.8	16
1965	55	4.1	2.7	19.5	23.6	14
1966	65	4.9	2.7	19.1	24.0	20
1967	75	5.6	2.7	18.3	22.9	24
1968	86	6.4	2.6	17.7	24.1	27
1969	96	7.2	2.4	17.5	24.7	29
1970	106	7.9	2.2	16.4	24.3	33
1971	119	8.7	2.1	15.6	24.4	36
1972	136	10.2	2.1	14.9	25.1	40

Source: State Institute of Statistics, Statistical Yearbook of Turkey (1973).

the danger of wind erosion compared with the tractor-pulled moldboard plow. Introduced as part of a properly tested package, however, tractor mechanization can be very useful, and even necessary, for achieving maximum agricultural growth. For example there are indications in Turkey that tractor mechanization could be an important part of a production package that would also include the following: new high-yield grain varieties; improved water conservation and water use practices; the use of more fertilizer; proper seed and fertilizer placement; and effective weed control. However, it will require extensive research to develop such a package and test it locally. On the other hand even doubling the number of tractors would not in itself initiate an agricultural revolution.

The First Five-Year Plan gave priority to the introduction of technically desirable machinery that would not have a labor-saving effect on a substantial scale. A decrease in the number of combines and a threefold increase in the number of animal-drawn grain drills were indicated as desirable. The number of combines has, however, continued to increase, and the increase in the number of grain drills, which can easily be manufactured in Turkey, has been relatively slow (see Table 5.14).

Mechanical stationary threshers, which could reduce the considerable breakage and waste of grain now experienced with the traditional threshing method, could be very useful. In 1971 there were about 19,300 threshers in Turkey and about 8,700 combines. In recent years the number of threshers has increased relatively more rapidly than the number of combines.

It is noted in the Second Five-Year Plan that the trend toward mechanization cannot be checked, since the need for planned, selective, and guided mechanization in Turkey does not seem to have been fully recognized. It appears that there is not a single research institution in Turkey working exclusively on the development and testing of agricultural tools and machinery. However, an agricultural machinery department has been established at the Ankara Institute of Agricultural Research.

Currently in Turkey tractors are mainly used in cotton and commercial wheat production areas, where relatively large operational land units exist. The distribution of tractors among the various regions of Turkey in 1968 is shown in Table 5.15. The data in the table indicate that the eastern part of the country, including the Black Sea, Central East, northeast, and southeast regions, are much less mechanized than the other parts of the country.

TABLE 5.14

Farm Mechanization Trends in Turkey, Selected Years

	Number of Machines								
	1948	1955	1960	1965	1968	1969	1970	1971	1972
Tractors	1,756	40,282	42,136	54,668	85,475	96,407	105,865	118,525	135,726
Tractor plows	1,119	28,840	31,528	42,600	63,700	82,000	87,017	98,284	109,463
Disk tractor plows	350	13,040	15,552	19,100	26,200	37,300	38,223	36,273	37,185
Disk harrows	680	16,150	18,268	22,480	32,200	36,700	41,613	54,442	53,593
Tractor cultivators	400	5,700	6,134	9,800	21,100	21,900	20,399	30,486	26,893
Grain drills (horse-drawn)	11,590	17,660	26,624	36,700	35,400	40,900	40,168	37,987	38,567
Grain drills (tractor-drawn)	160	7,240	8,343	12,700	15,200	24,700	24,600	19,390	21,308
Mowers	14,380	20,710	24,245	26,300	25,000	26,200	31,120	23,749	25,137
Binders	330	1,650	3,037	4,500	4,200	3,900	3,990	4,308	3,965
Combines	270	5,620	5,554	6,540	8,200	8,300	8,568	8,662	9,029
Threshers	430	1,329	2,536	4,300	12,000	11,400	14,044	19,312	24,485

Source: State Institute of Statistics, Statistical Yearbook of Turkey (1973).

This lack of mechanization in the eastern part of the country corresponds with the area's general backwardness. In the Black Sea region the fact that the farms are small also limits tractor mechanization. Of all the provinces Adana had the largest number of tractors in 1968, and nearly 9 percent of the total number in the country.

TABLE 5.15

Geographic Distribution of Tractors in Turkey, 1968

Coastal Regions	Number of Tractors	Inland Regions	Number of Tractors
Black Sea	3,002	Central North	13,415
Marmara	13,864	Central South	10,455
Aegean	18,646	Central East	6,351
Mediterranean	14,235	Northeast	1,837
		Southeast	3,670

Source: State Institute of Statistics, Turkey, Summary of Agricultural Statistics (1968).

Traditional agricultural technology still prevails in Turkey to a large degree. Almost all of the inter-plant tillage is done by hand hoeing, nearly all field crops are harvested by hand, and the threshing sled still dominates.

Because of the surplus of manpower in Turkish agriculture, tractor mechanization should occur gradually and in selected regions. An improvement in the quality of draft animals, along with the design and production of more efficient and effective animal-drawn equipment, also deserves attention. At the same time, seed-fertilizer drills and mechanical threshers may deserve rapid and widespread introduction.

AGRICULTURAL CREDIT

The increased use in Turkey of agricultural inputs such as fertilizers, improved seeds, and tractors has required substantial investments by the farmers, and since most farmers have not accumulated sufficient funds to finance these inputs, they have had to buy extensively on credit: approximately 80 percent of all fertilizer is sold on credit.

The private groups and government agencies that provide credit to the agricultural sector in Turkey can be classified under two headings: (1) Unorganized credit facilities, consisting mainly of individuals such as dealers and village money-lenders (credits acquired from such sources usually have very high interest rates); (2) Organized agricultural credit agencies: the Agricultural Bank, agricultural credit cooperatives, and agricultural sales cooperatives.

Actually the three organized sources of credit are interrelated organizations. The cooperatives are completely under the supervision of the Agricultural Bank. Therefore, as previously noted, the Agricultural Bank can be said to be the only actual organization that extends credit to the farmers.

With its present resources the Agricultural Bank is not in a position to meet all the credit needs of the Turkish farmers. Consequently it appears that in general from 30 to 50 percent of the credit (depending on the region) is provided by noninstitutional sources.

The Agricultural Bank offers three types of credit to the farmers: short-term credit, medium-term credit, and long-term credit. Table 5.16 shows the distribution of agricultural credit provided by the Agricultural Bank for the period from 1965 to 1969, according to the areas of agricultural development that the credit was used for.

Since the Agricultural Bank is the only government agency that extends credit to the farmers, and since the organizational rules and regulations of the bank decree that the credit needs of the small farms be given priority, one would expect the credit situation for the small farms to be relatively good. Unfortunately this is not true, and only a very small portion of the need for credit is met. Many poor farmers reportedly spend the limited amount of money that they earn for purposes other than agricultural investment. For this reason it becomes increasingly difficult for these farmers to pay their debts, especially following a bad crop year.

There are many bottlenecks in the provision, distribution, and repayment of agricultural credit in Turkey. The major problems are:

1. The credit allocated to individual farmers to meet their farming needs is insufficient.
2. The requests to farmers to make payments on the credit allocated to them are poorly timed.
3. Farmers misuse credit, basically failing to use it for agricultural purposes.

4. Farmers have inadequate pay-back capabilities, sometimes caused by the fact that they tend to resell the seeds, fertilizers, and other inputs provided to them in kind, rather than using these items for their own production needs.
5. There is an insufficiency of total bank resources for agricultural credit.
6. Inefficient and time-consuming procedures delay the processing of farmers' credit applications to the Agricultural Bank.
7. Farmers have difficulty meeting the guarantee requirements.
8. A high rate of interest is charged on farm credit.

TABLE 5.16

Credit Given by the Agricultural Bank of Turkey for Various Farm Production and Development Uses, 1965-69 (million TL)

Type of Credit and Use	1965	1966	1967	1968	1969
Short term					
Water development	10	15	20	20	20
Artificial fertilizers	120	225	300	469	730
Livestock	130	180	220	300	340
Certified seeds	20	75	240	300	350
Priority export products	335	370	400	450	500
Crop protection chemicals	n.s.	n.s.	100	100	100
Others	445	465	501	588	600
Total	1,060	1,300	1,781	2,227	2,640
Medium and long term					
Farm machinery and equipment	305	426	501	611	750
Water development	23	33	33	33	33
Crop protection machinery and equipment	12	13	35	44	46
Livestock	75	130	170	254	255
Draft animals, carts, and equipment	5	5	5	5	5
Others	220	223	258	250	165
Total	640	830	1,002	1,197	1,254
Overall Total	1,700	2,160	2,783	3,424	3,894

Note: n.s. = Not significant amounts.

Source: Agricultural Bank.

The Agricultural Bank is often more security-oriented than it is production- or profit-oriented. Direct loans to farmers are made with their land serving as collateral, and the land must be surveyed or registered before it can be put up as loan collateral. Only about 30 percent of the farmers have their land registered, because registration takes time and money. As a result the large-scale, wealthy farmers are usually the ones eligible to borrow money directly from the Agricultural Bank. Loans for operating expenses are usually made on a yearly basis, with a large proportion of these provided through the credit cooperatives. If the farmer does not pay back the entire amount of his loan he is not eligible to borrow the following year. For example, in the fall of 1970 a sizable quantity of improved wheat seed and fertilizer were not purchased by growers in the rainfed areas because a poor crop year for 1969/70 prevented many farmers from repaying their loans. Not only were these farmers not eligible to borrow additional money, but the funds tied up by the unpaid loans reduced the amount held by the Agricultural Bank for loans to new borrowers.

To encourage farmers to use their loans more efficiently, attempts have been made by the Agricultural Bank to offer a supervised credit program. In the Denzili area a pilot program of this type was started with the assistance of AID. At the end of 1967 there were 3,900 farmers participating in the program, and the amount of credit distributed was about TL 67.8 million. The goal is to expand the program to reach 60,000 new farmers each year. So far the program has not been considered too successful, mainly because it lacks coordination with the extension activities of the Ministry of Agriculture and it lacks manpower with practical training.*

A second source of credit is provided by the bank through the large production and marketing cooperatives, such as the Sugar Company. This institution provides necessary credit at a reasonable rate of interest to farmers growing sugar beets, enabling them to adopt and carry out efficient sugar beet production practices. Much of this credit is extended in the form of seed, fertilizer, and insecticides. Technical assistance is also provided to the farmer by this organization.

---

*James I. Blolock, "Capital and Finance in Turkish Agriculture" (Ankara: Economic Planning Division, AID, May 1969).

The credit cooperatives are another important source of credit with a total capital of nearly TL 2.0 billion. Also under the control of the Agricultural Bank, these cooperatives give short-term loans (one year or less) to their members at reasonable interest rates (6 percent per annum plus 3 percent of the loan for the cooperative's building and reserve funds).

One of the greatest weaknesses of the credit cooperatives' loans is their slow turnover, or velocity, of funds. It is estimated that every year about 20 to 25 percent of their loans are not repaid. This means that a substantial part of the short-term money is immobile each year. One of the reasons for this is that the borrowed funds are partly used for personal consumptive purposes; other reasons are the roundabout procedure for obtaining the loans and the lack of incentives to repay them on time.*

The amount of a loan issued by the cooperatives is based on the value of the borrower's property, not on his repayment capability. Since many farmers have small holdings most of the loans received are too small. In 1966 as many as 330,000 loans, or more than 35 percent of the total, were smaller than TL 500 ($55), and 27,000 members of the cooperatives received loans totaling less than TL 100 ($11). These small loans contribute little to the adoption of new production technology. A consequence of the many small loans is that a number of cooperatives are operating at a loss. Administrative paper work is complicated and time-consuming, and the administrative costs involved in issuing loans are too high. This problem, along with the farmers' poor repayment record, the misuse of loans by farmers for consumptive uses, and the inadequacy of overall funds received from the Agricultural Bank (the main source of the cooperatives' funds) have caused these cooperatives to be somewhat ineffective in increasing the use of new and improved production technology.

The amount of noninstitutional credit supplied to farmers by merchants, village money-lenders, relatives, and friends is not known, but in some regions private credit is estimated to be twice that of institutional credit. It is almost certain that the interest rates for noninstitutional credit are high (40 to 50 percent),

---

*P. J. Van Dooren, "Organization and Operation of Agricultural Credit Marketing and Village Development Cooperatives," No. TA 2557 (Rome: FAO, 1968).

TABLE 5.17

Retail Credit Provided in Turkey by the Agricultural Bank, Direct and through Credit Cooperatives (thousand TL)

Year	Current Bank Loans	Percent	Overdue	Percent	Bad Debts	Percent	Direct Loans	Loans by Cooperatives	Direct and by Cooperatives	Increase in Total New Loans
1962	894	n.r.	n.r.	n.r.	n.r.	n.r.	894	n.r.	894	n.r.
1963	17,273	75	5,011	22	715	3	23,000	n.r.	23,000	22,016
1964	27,612	48	27,332	48	2,357	4	57,301	22,888	80,189	57,189
1965	48,575	75	11,596	18	4,530	7	64,701	42,352	107,053	26,864
1966	65,018	71	19,405	21	7,683	8	92,106	48,352	140,458	33,405
1967	122,253	72	31,974	19	14,441	9	168,668	93,784	262,432	121,994
1968	165,211	61	77,842	28	29,264	11	272,317	243,896	516,213	253,771
1969	195,617	48	133,766	33	78,523	19	407,906	351,112	759,018	242,805
1970	112,417	28	161,908	41	124,703	31	399,028	395,906	794,934	35,916

Note: n.r. = Heading not relevant.

Source: John M. Hill, "The Need for Reorganization of Turkey's Fertilizer Industry" (AID: Ankara, February 2, 1971).

TABLE 5.18

Status of Selected Agricultural Bank Loans to Turkish Farmers as of December 31, 1969
(thousand TL)

Type of Credit and Use	Current	Percent	Overdue	Percent	Bad Debts	Percent	Total
Short term							
Fishery operation	2,951	50.5	2,164	37.0	731	12.5	5,846
Commercial fertilizer, retail	195,617	47.9	133,766	32.8	78,524	19.3	407,908
Commercial seed, retail	177,643	52.3	100,062	29.5	62,003	18.3	339,708
Livestock feed	209,534	74.8	47,026	16.8	23,572	8.4	280,132
Pest control chemicals	1,040	6.8	9,325	60.8	4,984	32.4	15,349
Marketing sales credit	37,567	87.5	3,487	8.1	1,863	4.4	42,916
Other operating credit	1,438,352	60.6	564,754	23.8	369,000	15.6	2,372,106
Medium and long term							
Fishery	8,829	55.8	3,980	25.1	3,019	19.1	15,828
Tractor credit	503,906	74.0	115,787	17.0	61,316	9.0	681,009
Other farm machinery	8,432	61.2	2,041	14.8	3,309	24.0	13,782
Processing plants and equipment	492	91.1	24	4.5	24	4.4	540
Pest control machinery	3,907	63.2	1,310	21.2	963	15.6	6,180
Livestock	130,239	69.5	37,705	20.1	19,517	10.4	187,462
Production improvement	11,074	39.9	7,475	26.9	9,205	33.2	27,754
Land purchase	217,787	77.6	35,814	12.9	26,693	9.5	280,294
Agricultural handcrafts	18,183	72.6	3,269	13.1	3,603	14.4	25,055
Total							4,702,520

Note: Numbers may not add due to rounding.

Source: John M. Hill, "The Need for Reorganization of Turkey's Fertilizer Industry" (AID: Ankara, February 2, 1971).

and that many of the loans are for the farmers' consumption rather than for agricultural development and production.

The amounts of credit provided by the Agricultural Bank in recent years, along with the repayment situation, is indicated in Table 5.17. The table indicates that repayments of retail fertilizer credit extended by the Agricultural Bank in recent years have become rather poor. This has been especially true in the wheat-growing areas of central and southeastern Turkey. In 1970 repayments ran about 36 percent on all fertilizer loans. Overdue and bad debts were 2½ times as large as current loans. Central and southeastern Turkey, which are the least developed parts of the country, had overdue and bad debts amounting to more than TL 200 million.

Table 5.18, giving data on the status of selected loans by the Agricultural Bank, shows that on loans for pest control chemicals the repayment situation is even worse than for fertilizer. It is also bad for credit on seed. For medium- and long-term credit the repayment situation is generally much better than for short-term loans.

Retail fertilizer credit accounts for approximately 8 percent of agricultural loans by institutional sources and for 14 percent of the credit cooperatives' loans. The credit extended for pest control is relatively insignificant.

CHAPTER

6

WHEAT MARKETING AND FLOOR PRICE SYSTEM

The government's intervention program for wheat in Turkey has the following main purposes: To reduce the price risk for the wheat producers, and provide them a reasonable return for their efforts in production; to ensure an adequate supply of wheat and bread at reasonable and stable prices for all Turkish consumers; to establish a margin (price spread) which will prevent exploitation by private sector merchants, but which at the same time will ensure that adequate private facilities are made available to assemble, store, and distribute wheat so that the program will not overburden the resources of the government.

The agency responsible for setting prices on wheat and other cereals is the Soils Products Office (Toprak Mahsulleri Offisi, or TMO)--a state economic enterprise which was established by law in 1938. According to the law the main functions of TMO are as follows:

1. To purchase wheat at officially determined prices and thereby prevent wheat prices received by farmers from falling below the normal level;
2. To provide stockpiles and, by selling wheat at officially determined prices, to prevent the price of wheat from rising to a level that would discourage consumers;
3. To import and export wheat;
4. To establish grading systems and price differentials in order to encourage the farmers to deliver better quality;
5. To own and operate flour mills, bakeries, etc.

There is a regional head office of TMO in each of the nine agricultural regions, and there are more than 300 TMO branch offices that purchase, store, and sell grains. In 1971 the responsibility of TMO was extended to include oil seeds, pulses, and feed grains. The agency was also given complete control over opium production and marketing.

TMO currently has a storage capacity of 1,750,000 metric tons scattered throughout Turkey--enough to store about 15 percent of Turkey's current annual agricultural production. This storage capacity was sufficient to allow orderly marketing of wheat in 1971 despite the high production for the year. The Central Bank provided the capital necessary to purchase all the wheat that was offered to TMO, and the technical know-how existed to procure and store the output in a relatively efficient manner.

Each year the director general of TMO recommends an appropriate TMO wheat price to the Ministry of Commerce and the Council of Ministers. In making his recommendations he considers the anticipated wheat production, market demand, and stocks. The Council of Ministers then sets the purchase (support) price at the recommended level, or higher, depending on the political climate at the time of its action and other economic information received. The purchase prices for the various types of wheat are announced at the beginning of the harvest or just prior to it--usually between May 1 and June 15. The main reason for this late announcement is the wide yield fluctuations for wheat. The government's price policy does, however, affect farmers' price expectations for the next year and future years.

The support price is uniform by type for specified qualities, and applies to wheat delivered to any one of the more than 300 procurement stations operated by TMO throughout the country. Essentially TMO announces that it will buy any quantity delivered at the specified purchase price. At the same time sales price for wheat applicable to municipalities and villages throughout the country is announced: from 1971 to 1972 this price was 10 kurus per kilogram above the purchase price. This applies to a specified quantity for each village or municipality.

TMO stores the wheat purchased and distributes it to private mills, municipalities, and villages on an "as needed" basis. A rationing procedure is used for allocation that provides a per capita supply of 10 kilograms per month. TMO delivers to each area, at the announced sale price, the difference between locally available supplies and the area's calculated need. Quantities in excess of the calculated need may be sold by TMO, based on its available supplies, at going wholesale market prices.

Both the average wholesale market prices and the reported average farm prices for wheat have normally been above the support price (see Figure 4). Except for 1971 TMO purchased less than 10 percent of the annual wheat production from 1957 to 1972 (see Table 6.1). The output purchased by TMO, with only two exceptions, ranged from about 470,000 metric tons to nearly 900,000 metric tons (see Table 6.2). This may amount to as much as 20 to 40 percent of commercial wheat sales, however, since a large percentage of wheat production is consumed where it is produced, and never enters the commercial market.

During years of short supply, TMO has imported wheat under the Public Law 480 program. TMO is the only authorized importer of wheat in Turkey. One of TMO's goals is to have approximately 300,000 metric tons of wheat on hand at the end of each marketing year (May 31). This goal has actually been achieved in six of the last 14 years. Current planning indicates that, in the coming years, TMO will have to buy and sell 1.2-1.4 million metric tons of wheat annually.

The balance of the commercial trading in wheat is carried out by the private sector: merchants, trucker collectors, large farmers, and flour mill operators. There is both a competitive and a complementary relationship between TMO and the private trade. In order to buy wheat reserves TMO can, if necessary, monopolize the transport facilities during the harvest season. However, it operates only one flour mill that enables it to gain practical experience and to establish market standards; thus it must rely on the services of local flour mill operators. TMO is a major factor in the distribution of wheat: for the 1968-72 period it has handled an average of more than 1.1 million metric tons per year.

Although TMO provides the means of importing or exporting quantities of wheat demanded in excess or short of available supplies, thus reducing price uncertainty at both the farm level and the consumer level, other uncertainties exist in Turkey's wheat industry. Commercial purchasers of wheat essentially follow the price leadership provided by TMO, and establish their price at a level just high enough above TMO's price to obtain their current and short-run needs. TMO establishes allocation quotas, and both the purchase price and the selling price, in advance. Therefore the millers know about how much wheat

FIGURE 4

TMO Wheat Support Price, Average Farm Price, and Wholesale Price in Turkey, 1957-72

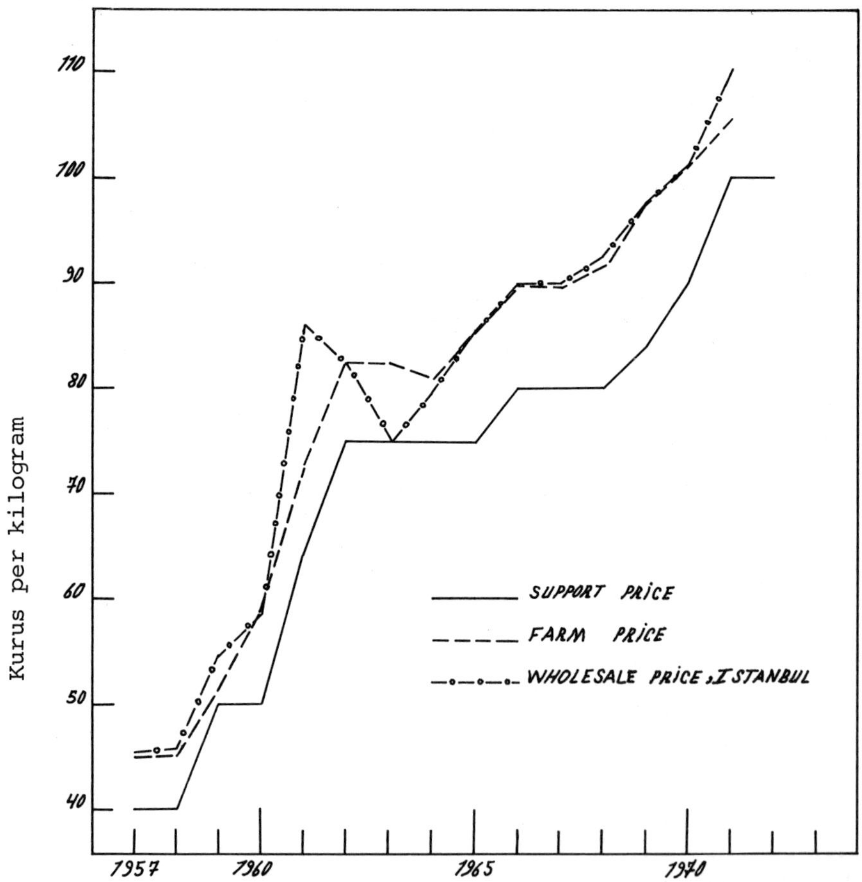

TABLE 6.1

TMO's Wheat Prices and Purchases Compared to the Turkish Market, 1957-72

Year	TMO Support Prices (kurus per kilogram)	Price Index for Wholesale (1963 = 100)	Deflated Support Price (kurus per kilogram)	Farm Price (kurus per kilogram)	Deflated Farm Price (1963 prices, kurus per kilogram)	Price Difference between Farm Price and Support Price (kurus per kilogram)	Price Difference in Percent of Support Price (kurus per kilogram)	Wholesale Price for Wheat (Soft) in Istanbul (kurus per kilogram)	TMO Purchases as Percent Total Production
1957	40	57.7	69.3	44.6	77.3	4.6	11.5	44.8	9.7
1958	40	59.4	67.3	44.9	75.6	4.9	12.3	45.9	9.3
1959	50	71.0	70.4	51.7	72.8	1.7	3.4	54.1	6.0
1960	50	77.3	64.7	59.1	76.5	9.1	18.2	58.3	4.5
1961	64	86.7	73.8	72.7	83.9	8.7	13.6	86.0	0.3
1962	75	96.9	77.4	82.4	85.0	7.4	9.9	82.5	5.7
1963	75	100.0	75.0	82.1	82.1	7.1	9.5	75.4	7.9
1964	75	96.2	78.0	81.4	84.6	6.4	8.5	79.3	4.5
1965	75	104.8	71.8	85.8	82.1	10.8	14.4	85.3	6.2
1966	80	105.4	76.0	89.9	85.3	9.9	12.4	89.5	8.3
1967	80	110.6	72.3	89.9	81.3	9.9	12.4	86.3	8.7
1968	80	113.7	70.4	91.7	80.7	11.7	14.6	92.2	5.3
1969	84	123.9	67.8	97.2	78.5	13.2	15.7	97.1	5.2
1970	90	127.0	70.9	100.8	78.7	17.0	18.9	101.1	8.1
1971	100	138.4	72.3	105.2	76.1	5.2	5.2	110.4	15.4
1972	100	153.4	65.1	109.8	71.4	9.8	9.8	124.5	n.a.

Notes: The annual figures in the table cover various periods. The support price covers the marketing year, June 1-May 31. The price index, the farm price, and the wholesale price cover the calendar year. n.a. = Data not available.

Sources: TMO; Ministry of Commerce; State Institute of Statistics; and Olan D. Forker, "Agricultural Price Policy in Turkey. A Collection of Works," vol. 2, Economic Staff Papers, Discussion Paper No. 3 (Ankara: USAID, 1971).

TABLE 6.2

Wheat Purchases and Sales by TMO in Turkey, 1957-58 to 1971-72 (thousand metric tons)

Year[a]	Purchases	Imports[b]	Sales	Exports	Stocks[c]
1957-58	738	311	738	n.s.	400
1958-59	797	128	724	319	315
1959-60	469	300	929	104	67
1960-61	386	373	753	1	72
1961-62	32	1,286	1,190	n.s.	193
1962-63	485	583	1,110	n.s.	129
1963-64	791	371	819	n.s.	508
1964-65	369	276	787	n.s.	335
1965-66	525	163	832	n.s.	144
1966-67	793	308	841	n.s.	357
1967-68	880	36	998	n.s.	323
1968-69	549	236	1,161	n.s.	99
1969-70	500	770	1,188	n.s.	118
1970-71	808	622	1,225	n.s.	268
1971-72	1,938	346	1,143	478	1,462

[a]Marketing year, June 1-May 31.
[b]Imports are primarily PL 480 purchases.
[c]Stocks are end-of-marketing-year figures per May 31.

Note: n.s. = Not significant amount.

Sources: TMO; and Olan D. Forker, "Agricultural Price Policy in Turkey. A Collection of Works," vol. 2, Economic Staff Papers, Discussion Paper No. 3 (AID: Ankara, 1971).

they will have to purchase from other sources and approximately how much more they will have to pay to get it. The close correlation between TMO prices and the free market prices exists mainly because of the following factors: (1) TMO is ready to purchase all of the wheat that is offered at the support price, thus providing an effective floor under which the farm price cannot go; (2) TMO is able (if necessary through imports) to supply the commercial market with a major part of its needs at the TMO fixed sales price, thus providing an effective ceiling for the free wholesale price of wheat; and (3) the government fixes the TMO margin at a level which will not fully cover costs under all circumstances. Because of this last factor amount, which can be considered as the approximate government net cost of intervention, the private merchants cannot take over more than that part of the market where they can compete at the relatively fixed margin.

The quantity of wheat that TMO purchases each year is a function of the level of production in relation to demand, of the price support level, and of the level of TMO stocks at the beginning of the year. The most important factor of course is the relative level of production.

Since the market price fluctuates at a level somewhat above the support price, and since TMO does not have large excess supplies at the end of the year, it appears that the price support policy for wheat is reasonably efficient in providing a floor for the farm prices. The TMO support price for wheat has usually been kept at the same level for only two to three years; it is then increased. In recent years the support price for wheat has increased less than the wholesale price index for vegetable foodstuffs, and as a consequence the real wheat support price has gone down approximately 14 percent. (See Table 6.1 and the general wholesale price index Statistical Appendix, Tables A.20 and A.21.)

The average annual increase in the outstanding credit from the Central Bank needed to finance TMO's purchases of wheat in the 1960s was TL 109 million. This represents only 1.4 percent of the total annual value of the wheat crop, and it appears that wheat, of the major crops, gets relatively the least support through government financing.

The current margin of 10 kurus between the purchase price and the selling price for all market locations represents a tendency to underprice the government services in the procurement, transportation, and storage of wheat. TMO incurs particularly high transportation costs serving all the remote wheat deficit areas. How the present margin corresponds to costs is indicated in Table 6.3.

TABLE 6.3

Hypothetical Costs of Marketing Wheat Produced in the Iskenderun Region of Turkey and Marketed in Iskenderun or Istanbul, 1971
(kurus per kilogram)

Cost Item	Iskenderun	Istanbul
Costs within region		
Local transportation	2	n.r.
Facilities and personnel	5	n.r.
Transfer costs to Istanbul		
Transportation	n.r.	6
Unloading	n.r.	2
Total cost at harvest including ownership cost of storage facilities	7	15
Total cumulative cost including cost of credit and cost of storage, calculated at 1 kuru per kilogram per month		
August	7	15
September	8	16
October	9	17
November	10	18
December	11	19
January	12	20
February	13	21
March	14	22
April	15	23
May	16	24

Notes: Costs are based on a technically efficient marketing system. n.r. = Heading not relevant.

Source: Olan D. Forker, "Marketing, Subsidies and Price Policy" (Paper delivered at October 1972 conference of the Economic and Social Studies Conference Board, Istanbul).

The table demonstrates the nature of TMO's losses. The flat procurement price and selling price regardless of the point of origin or point of sale or time of the season represents an averaging out of TMO's real costs. One result of this is that the private sector will provide most of its marketing services in particular transportation situations, and seasons, where it has a likely profit to gain. TMO activities for all practical purposes are restricted to those areas where the private sector would not expect to make a profit.

The TMO loss will depend on the size of the fixed margin. The wider the margin, the more involvement one would expect from the private sector; the narrower the margin, the less involvement one would expect from the private sector. TMO, however, needs a certain part of the market to achieve its objectives. This is the logical explanation for TMO's establishing a margin below average cost, and thereby taking a loss each year in its operations.

The system also represents an important social policy to the more remote surplus and deficit areas, which also often have relatively low per capita income. Different buying and selling prices for different markets would hit the less developed areas, and the poor producing and consuming areas, especially hard.

The TMO loss could be partly offset by instituting a seasonal price differential to allow for storage costs during the year. Such a policy would encourage more involvement from the private sector during the later parts of the season following the wheat harvest. A seasonal variation in the selling price for wheat would, however, make the stabilization of bread prices more difficult.

Wheat bread is the most important food item in the Turkish diet. In most municipalities the price of bread and the weight of the loaf are fixed by the municipal government, according to the price of wheat fixed by the federal government. Although there is some private movement of wheat and wheat flour, the effective cost of wheat and wheat flour to millers and bakers is closely related to the government's selling price. Studies show that the farmers' share of the retail value of bread is between 56 percent and 66 percent. From 34 percent to 44 percent is absorbed by transportation, storage, milling, and baking. If the current margins are adequate to cover costs and provide a reasonable return on investment, a 10-kurus increase in wheat prices would have to be offset by a correspondingly smaller increase in the price of bread.

TMO also purchases and sells barley, rye, oats, and maize. The procedure used for these grains is similar to the procedure used for wheat except that TMO does not attempt to exert as much price influence as with wheat. However, purchases of these grains by TMO have been as high as 6 percent of total production. The agency also acts as an importer of maize and an exporter of barley. Average farm prices for both barley and rye have generally been above the stated support prices.

Since the government subsidy of 1.4 percent is rather insignificant, there might appear to be a conflict between TMO's dual function to protect the interest of the producers by maintaining a fair price level and to defend the interest of the consumers by preventing selling prices from rising too high. This makes it necessary to strike a balance. Since the Turkish support price usually has been above the world market price, TMO has been able to enjoy substantial markups as a result of its import monopoly. The present Turkish support price of TL 1,000 per ton is equivalent to $71.43 per ton, at the 1972 rate of 14 Turkish lira to 1 U.S. dollar: this may be characterized as a moderate price support level.

Despite its problems, TMO has done a relatively effective job compared with similar marketing arrangements in other countries and compared with other alternatives. It appears to be preferable to have a single agency, as in Turkey, deal with the whole complex of problems regarding a wheat marketing and floor price system, rather than to split up the main functions between several ministries.

CHAPTER

7

INCENTIVE POLICIES AND GOVERNMENT INVOLVEMENT

Economic incentive, which is an essential condition for agricultural development and increased agricultural output, can take several forms. One is to make sure that all farmers have access to market outlets. Another is to establish a favorable relationship between the prices that farmers receive for their products and the prices they pay for production inputs. Price supports on products sold by farmers or subsidies on the inputs used in production are important means for influencing this input-output price relationship. A third economic incentive is direct or indirect assistance given to the marketing process to assure price stability and the efficient operation and development of the marketing system. A fourth type of economic incentive is that which results from reduced taxes or fees affecting the farmers' production capacity. Fifth, sharecroppers must be assured a reasonable share of the annual crop. Finally, the availability of goods and services that farmers purchase can serve as an incentive to increased production.

The current economic incentive policies in Turkey have four goals, as pointed out by Olan Forker: to expand total production of food and fiber to or beyond the level of self-sufficiency; to use agricultural output to produce a favorable balance of trade; to increase farm prices; and to maintain lower consumer prices.*

---

*Olan D. Forker, "Agricultural Price Intervention by the Government of Turkey," Paper presented at the USAID Turkey conference in August 1967; and Olan D. Forker, "Marketing, Subsidies and Price Policy," Paper presented at the October 1972 conference of the Economic and Social Studies Conference Board, Istanbul.

The responsibility for carrying out programs to meet these goals is vested primarily in the Ministry of Commerce. The ministry uses four general kinds of price intervention programs: the direct manipulation of commodity prices; the subsidization of production inputs; the establishment of state economic enterprises and sales cooperatives to provide marketing services and to control sales; and the exertion of direct influence on export prices.

The ministry's intervention through the control or the influencing of the prices of agricultural products has been considerable. For hazelnuts, raisins, and figs, and, more recently, for cotton, olives, and pistachio nuts, this intervention has been direct. Funds are provided to sales cooperatives to set floor or support prices through government purchase and storage. For cereals, sugar, and meat the government has influenced prices through its state economic enterprises, using purchase programs and contracts. In the cases of tobacco and tea, price influence is through the state monopolies. The monopolies purchase a portion of production adequate for the country's domestic needs, and market the commodity both domestically and abroad.

For all of these commodities, except sugar, the price is announced at the beginning of the harvest, or just prior to it, when supplies are fixed. In his paper on agricultural price intervention of 1962 Forker points out that in many years inappropriate production estimates have resulted in the accumulation of excessive stocks by the various intervening agencies, in some cases increasing product deterioration and loss.

Minimum export prices were established for many products until 1965, when this system was abolished and all export sales began to be licensed. As a result export sales can now be controlled by the export trade unions. The unions establish minimum prices consistent with Turkish and world supply and demand conditions. The state economic enterprises, the state monopolies, and the sales cooperatives must be members of these unions.

In his paper on agricultural price intervention of 1967 Forker makes the following general comments on the impact of these price interventions:

> The motivating forces behind the intervention programs and the established price levels are both economic and political, but mostly political.
>
> Wheat production has probably been held below potential levels through the

establishment of price levels that concurrently permit the importation of wheat each year that can be moved into the domestic market, but this action in itself has also held consumer prices at lower levels than they would be if the imports were limited.

Raisins and hazelnuts production has expanded at a faster rate than has the market (domestic plus export). The current level of prices evidently encourage continued expansion, subsidization of producer income and burdensome carry-over stocks.

Tobacco production has also expanded at a faster rate than has the market at the intervention price levels. It appears that the monopoly practice of pricing a high quality high cost tobacco at a higher level than the world market will bear and the low quality at a lower price than the market will bear are encouraging the exportation of lower quality Turkish (oriental) tobacco. If this is true than this is contrary to the goal of maximizing foreign exchange earnings.

Sugar production exceeds domestic needs at established prices and evidently some subsidization takes place. However, the mechanics are available for controlling production and manipulating price levels. The evident policy is one of the subsidization of sugar beet producers and sugar consumers.

Subsidization of the domestic market system and thus producers and consumers take place in wheat (storage and transportation), meat (processing, cold storage and distribution), sugar and tobacco.

Any intervention program involves the use of government resources. In Turkey most of the cost is borne by the Central Bank. Credit is extended to the state enterprises, the state monopolies, or the sales cooperatives to provide money for purchases. Losses incurred by these agencies in implementing the intervention policy accumulate as outstanding credit to the agency. The average increase in credit extended for this purpose over a period of years can be considered as the approximate government net cost of intervention. The total cost is higher than this to the extent that general funds (revenue) are used to support the operating costs of the government enterprises. Table 7.1 gives the approximate average annual costs of intervention for the 1950-60 period and the 1960-70 period, according to Forker.

TABLE 7.1

Average Annual Increase in the Outstanding Credit of the Central Bank to Agencies Involved in Agricultural Price and Market Intervention Activities (million TL per year)

Agency and Commodity Purchased	1950-60	1960-70
Soil Products Office (wheat)	80	109
Sugar Factories Company (sugar)	41	62
Monopoly-administered (tobacco and tea)	84	175
Tobacco financing	10	15
Sales cooperatives	53	129
Total increase in credit for intervention per year	208	490
Increase in total current credit	520	1,371
Intervention as percentage of agencies' total operating costs	40	36

Source: Olan D. Forker, "Marketing Subsidies and Price Policy" (Paper presented at the October 1972 conference of the Economic and Social Studies Conference Board, Istanbul).

The average annual cost for the 1950-60 period was about TL 208 million. For the 1960-70 period the annual cost amounted to about TL 490 million. During the 1950s intervention in wheat prices and marketing incurred the largest cost, an average of TL 80 million per year. But during the 1960s the losses incurred on tea and tobacco intervention accounted for the largest cost, amounting to about TL 190 million per year. The total cost of intervention is equivalent to about 2 percent of the national income of the agricultural sector. If one were to add the direct subsidy to production through fertilizer, seeds, and water (the water charges cover only operation and maintenance costs), intervention costs would be equivalent to almost 2.5 percent of the national income from the agricultural sector.

However, the relative importance of the government's involvement varies considerably among the major commodities, as indicated in Table 7.2.

TABLE 7.2

Importance of Implied Cost of Government Intervention in Turkish Agriculture Relative to the Value of the Commodity, 1960-70

Commodity	Average Annual Cost (million TL)	Implied Cost as a Percent of Total Value of Crop
Tobacco and tea	180.0	12.4
Sugar	62.0	11.6
Hazelnuts	58.6	9.2
Raisins	9.3	6.5
Cotton	78.3	4.2
Olive oil	16.4	2.6
Pistachios	2.6	2.2
Figs	2.9	1.5
Wheat	109.0	1.4

Source: Olan D. Forker, "Marketing Subsidies and Price Policy" (Paper presented at the October 1972 conference of the Economic and Social Studies Conference Board, Istanbul).

The relative cost of intervention in sugar and in tobacco and tea is more than 10 percent of the total value of the crop, while for wheat the cost is only 1.4 percent. Compared to other major commodities wheat actually has received the least support calculated as percentage of the total value of the crop.

Several types of subsidies for production inputs are used by the Turkish government. One type takes the form of sales of nitrogen fertilizers at the c.i.f. import price. This is below the cost of Turkish production. In reality this may be considered a subsidy only in relation to the cost of locally produced fertilizers. On the other hand it can be viewed as a subsidy to the fertilizer-producing operations. However, the Turkish capacity to increase fertilizer production is increasing yearly and further efficiencies and lower internal unit costs due to near-capacity production should materialize in the near future.

Sales of seeds to the farmers by the government at the same price as for food or feed have been common. Also farmers have limited free or low-cost use of agricultural machinery, which is owned by the government. Interest rates for agricultural credit have been low, and only have in some cases covered the administrative cost of the government's credit agency.

Crop production services have been free in those areas where they are available. Particularly for sugar, but also for tobacco and some other products handled by government-controlled enterprises, indirect subsidies to farmers in the form of technical advice on farm problems and production assistance have been effective in helping to bring about modernization of production.

Price ceilings on agricultural products, which usually serve as a disincentive to agricultural production and modernization, do not generally exist in Turkey. On the other hand the consuming public has been subsidized through the government's sale of wheat to millers at below-cost prices.

The extent of the government's involvement in the procurement and marketing of agricultural products, and in the pricing, manufacturing, and distribution of agricultural inputs, is indicated in Table 7.3 and Table 7.4, respectively.

The modernization and increased efficiency of the marketing system can be an important means of increasing farm income. With the assistance of the AID study team, the World Bank, and other international agencies, attempts are underway to modernize and develop the processing and marketing facilities for fruits, vegetables, dairy products, and meat products.

Land taxes in Turkey are moderately low. Tax assessments are based upon value appraisals that were made in the 1930s, when land values were only a fraction of their current level. Thus land taxes do not appear to be an important disincentive to agricultural production and development.

It is reported that few farmers pay income taxes, and their payments are small. Farmers do indirectly pay an important part of the excise taxes applicable to the general public because of the size of the agricultural

TABLE 7.3

Government Involvement in the Procurement and Marketing of Agricultural Products in Turkey
(X indicates involvement)

Commodity	Government Supports Purchase Price	Procurement from Farmer: SEE[a]	Procurement from Farmer: Cooperatives	Procurement from Farmer: Private	Domestic Sales: SEE[a]	Domestic Sales: Cooperatives	Domestic Sales: Private	Municipality Sets Ceiling Prices for Consumers	Export/Import[b]: SEE[a] and Cooperatives	Export/Import[b]: Private
Wheat	X	X	n.r.	X	X	n.r.	X	X bread	X	n.r.
Other cereals	X	X	n.r.	X	X	n.r.	X	n.r.	X	n.r.
Pulses	X[e]	n.r.	n.r.	X	X[e]	n.r.	X	n.r.	X[e]	X
Tobacco[c]	X	X	n.r.	X	X	n.r.	n.r.	n.r.	X	X
Potatoes	n.r.	n.r.	n.r.	X	n.r.	n.r.	X	n.r.	n.r.	X
Sugar beets[d]	X	X	n.r.	n.r.	X	n.r.	n.r.	n.r.	X	n.r.
Cotton	X	n.r.	X	X	n.r.	X	X	n.r.	X	X
Oilseeds	X[e]	n.r.[e]	X	X	n.r.[e]	X	X	n.r.	n.r.[e]	X
Opium gum	X	X	n.r.	n.r.[f]	X	n.r.	n.r.[f]	n.r.	X	n.r.[f]
Vegetables	n.r.	n.r.[g]	n.r.	X	n.r.	n.r.	X	n.r.	n.r.[g]	X
Fruits	n.r.	n.r.[g]	n.r.	X	n.r.	n.r.	X	n.r.	n.r.[g]	X
Grapes	n.r.	n.r.[g]	X	X	n.r.	X	X	n.r.	n.r.[g]	X
Citrus fruit	n.r.	n.r.[g]	X	X	n.r.	X	X	n.r.	n.r.[g]	X
Other	n.r.	n.r.	X	X	n.r.	X	X	n.r.	n.r.	X
Hazelnuts	X	n.r.	X	X	n.r.	X	X	n.r.	X	X
Pistachios	X	n.r.	X	X	n.r.	X	X	n.r.	X	X
Dried figs	X	n.r.	X	X	n.r.	X	X	n.r.	X	X
Raisins	X	n.r.	X	X	n.r.	X	X	n.r.	X	X
Olive oil	X	n.r.	X	X	n.r.	X	X	n.r.	X	X
Livestock	n.r.	X	n.r.	X	X	n.r.	X	n.r.	X	X
Meat	n.r.	n.r.	n.r.	n.r.	X	n.r.	X	X	X	X
Wool	n.r.	n.r.	n.r.	X	n.r.	n.r.	X	n.r.	n.r.	X
Tea[c]	X	X	n.r.	n.r.	X	n.r.	n.r.	n.r.	X	n.r.

[a]State Economic Enterprises.

[b]All exports and imports are subject to some degree of control. A license must be obtained from the appropriate government agency.

[c]Tobacco and tea purchases and sales are made by the state monopoly.

[d]The Sugar Factories Corporation is a state economic enterprise that has the exclusive right to purchase, sell, and export and import sugar beets and sugar.

[e]It is planned that the Soil Products Office will expand activities in pulses, oilseeds, and livestock feed.

[f]Illegal private-sector trade does probably exist.

[g]Procurement for export has been made by MEPA, a state-supported enterprise, since 1969.

Note: n.r. = Heading not relevant.

Source: Olan D. Forker, "Marketing Subsidies and Price Policy" (Paper presented at the October 1972 conference of the Economic and Social Studies Conference Board, Istanbul).

TABLE 7.4

Government Involvement in the Pricing, Manufacturing, and Distribution of Agricultural Inputs in Turkey
(X indicates involvement)

Input	Government Sets Selling Price	Production or Manufacture		Distribution		Imports[a]	
		Public	Private	Public	Private	Public	Private
Fertilizer	X	X	X	X	X[b]	X	X[b]
Certified seeds	n.r.	X	X	X	X	X	n.r.
Pesticides	n.r.	n.r.	X	X	X	n.r.	X
Tractors	X	X	X	X	X	X	X
Combines	X	n.r.	n.r.	X	X	n.r.	X
Other farm machinery	n.r.	n.r.	X	X	X	n.r.	X
Feeds	n.r.	X	X	X	X	n.r.	X

[a]All imports are subject to some degree of control. A license must be obtained from the appropriate government agency: the purpose is to control the flow of currency.

[b]Until 1971 private sector was permitted to import and distribute compound fertilizer and potash fertilizer.

Note: n.r. = Heading not relevant.

Source: Olan D. Forker, "Marketing Subsidies and Price Policy" (Paper presented at the October 1972 conference of the Economic and Social Studies Conference Board, Istanbul).

sector. However, it can be generally concluded that there is no important disincentive to the agricultural sector resulting from taxes and fees.

The unavailability of certain consumer goods that farm families wish to buy may serve as a disincentive to increasing incomes in some areas of Turkey. Because of the rather strict import policy and the shortage of foreign exchange, a variety of modern consumer goods ranging from transistor radios to automobiles have been rather costly. However, this has been partly offset by the increased availability of improved agricultural production equipment and consumer goods produced locally. In addition basic foods are available in quantity. On the other hand the prices of many goods have been increasing at a faster rate than prices for farm products.

It can be concluded that the economic incentives for agricultural growth in Turkey appear to have been moderately favorable during the past few years: Prices have been subsidized; seasonal and yearly variations in price have been somewhat stabilized; certain input costs have been reduced; marketing efficiencies have been improved; and no serious disincentives exist as a result of taxes, fees, or the availability of consumer goods. The future may present a different situation, and serious study of the most effective types and degrees of incentives are needed if future agricultural development and output are to be maximized.

CHAPTER

8

THE AGRICULTURAL DEVELOPMENT SERVICES

AGRICULTURAL FIELD SERVICES

The agricultural field services maintained in Turkey are based on the administrative structure of the various provinces, which, like local administration throughout much of the Middle East, is controlled from the capital, in this case Ankara. All major social and economic functions are performed at the provincial level by officials of the central ministries. The provincial governor has a key position in this administrative structure. He issues instructions to government offices within his province, including the agricultural field services. As a result the extension staff spends more than 50 percent of its time on nonagricultural duties. Thus government agencies operate at the village level with little or no coordination and with low effectiveness.

The functions of the six agencies of the Ministry of Agriculture that have direct contact with the villages are the following:

1. Technical Agricultural Organization: This agency is under the General Directorate of Agriculture. It is the largest of the services and represents essentially an agricultural extension service. It has a technical agricultural director in each province, and has a graduate-level staff in each district.

2. General Directorate of Veterinary Services: This organization has a regional and provincial organization with veterinary officers and animal health assistants at the district level. The personnel are primarily concerned with disease control, and to a lesser degree with extension

work among farmers in livestock improvement, feeding, and health. Improved breeds are also made available to farmers from the directorate's stud farms.

3. General Directorate of Plant Protection and Quarantine: This agency also has a regional and provincial organization. Its functions at the district level are carried out by a technical agricultural staff.

4. General Directorate of State Farms: There are 22 state farms. Their main function is the multiplication, and distribution to farmers, of improved cereal seeds and livestock. Also they serve as a production demonstration arm of the ministry.

5. Agricultural schools: Some of these run short courses for farmers, and a few have an agent attached to them on an experimental basis who carries out extension work in surrounding villages.

6. Research stations: Some of these have demonstrations, field days, and short courses for farmers, and undertake certain extension activities in specialized cash crops and seed production.

The more important agricultural field services of the other government ministries that are active in the villages are as follows:

1. Ministry of Village Affairs: This ministry's General Directorate of Soil and Water Conservation (Toprak Su) is primarily concerned with the utilization of water in both the large irrigation programs and the many smaller irrigated areas. It has a large professional staff and an intensive program of farmer training. Another agency of this ministry, the Department of Forest Villages, in the General Directorate of Organizations and Credit, has a relatively small staff for administration and credit programs, and development work in the forest villages, which represent 37 percent of the total number of villages.

2. Ministry of Forestry: This ministry is also active in development work in forest villages, and the full responsibility for forest village development may be transferred to this ministry.

3. Ministry of Commerce: This ministry has two important agencies working in the field. First, there is the Agricultural Bank, which operates supervised agricultural credit programs in about 20 provinces, and has plans to extend these facilities to all the provinces. Agricultural staff are attached to the bank's headquarters and provincial offices for this work. Second, there is

the Turkish Sugar Factories Corporation, which has an agricultural staff responsible for closely supervising and offering advisory assistance to all sugar beet growers under contract to each of its 17 factories. The agricultural field work is supported by a number of experimental farms. The beet growers in fact are controlled by a contract drawn up between the corporation and groups of farmers from each village.

All of these technical services are highly centralized, with little cooperation or coordination occurring at the provincial and local levels, as noted above. The numbers of technical personnel of the main agencies functioning at the provincial and local levels are given in Table 8.1.

TABLE 8.1

Technical Personnel of the Main Agricultural Service Organization in Turkey at the Provincial and Local Levels, November 1969

	Number of Technical Agricultural Personnel		
Organization	University Graduates	Technicians*	Total Personnel
Technical Agricultural Organization	444	2,396	2,840
Veterinary Services	726	898	1,624
Plant Protection and Quarantine	203	399	602
Soil and Water Conservation	567	10	577
Agricultural Bank	60	81	141
Sugar Factories Corporation	192	194	386
Total	2,192	3,978	6,170

*Mostly with upper secondary-level training.

Source: State Planning Organization, Turkey.

The Technical Agricultural Organization (of the Ministry of Agriculture) carries the primary responsibility for agricultural extension. However, it has not been given sufficient personnel and material resources to carry out either the extension function or the many regulatory and administrative activities that are required. A proper operational and organizational framework for cooperation with agricultural research is also lacking. The staff is insufficient, with each extension worker often assigned to more than twice the desirable maximum number of farm families. The lack of transportation is a major handicap, along with the shortage of office facilities and clerical staff.

A major part of staff time is devoted to statutory and regulatory functions. Extension work per se is given very little time. Therefore it is questionable whether an agricultural extension service, in the real sense of the term, exists in Turkey at the present time.

AGRICULTURAL RESEARCH AND DEVELOPMENT SERVICES

A number of agencies provide agricultural research and development services, with the General Directorate of Agriculture of the Ministry of Agriculture serving as the chief agency in this regard. Various crop research committees meet once or twice a year, and submit reports to the National Crop Advisory Committee, which is comprised of representatives of the other general directorates of the Ministry of Agriculture, and representatives of the other ministries, the Agricultural Bank, universities, and Turkish agricultural chambers. The National Advisory Committee discusses general policy matters and advises on projects prepared by any of the participating agencies. The crop research committees have only advisory status and have been mainly concerned with cereals.

The National Scientific Research Council also has an Agricultural Group, with a secretariat, in Ankara. The group submitted, some years ago, a report to the Ministry of Agriculture regarding the reorganization of agricultural research. It has its own funds and allocates them directly to individual projects, mainly to staffing the agricultural faculties at the universities. The Agricultural Group has considered starting its own dry land agricultural research institute.

The 20 agencies involved in agricultural research and development, and their activities, are as follows:

1. General Directorate of Agriculture (Ministry of Agriculture): This agency has four subdepartments, and, as noted above, is the main agency. Divisions within these departments are mainly on a subject-matter basis. For example, the research and production department is responsible for the major portion of agricultural research. It operates 75 institutions for research and development, and employs over 400 professionals and 100 technicians. Many of the institutes are small with only a few employees. Most of them have sufficient land; inadequate operational budgets and a shortage of properly qualified staff are the main constraints. There are far too many specialized stations. As a result the few good research workers are thinly spread, and partly occupied by administrative matters. Effective coordination is missing, and often the staff works in isolation.

2. General Directorate of Veterinary Services (Ministry of Agriculture): This directorate is responsible for a wide range of research, development, and control activities, including those directly affecting animal health and diseases; those concerned with the control of pharmaceutical products, food, and feed for animals; and those concerned with the breeding and improvement of farm animals. About 280 in graduate staff are employed at the directorate's various institutes, laboratories, stud farms, and so on. Only 164 of these are principally research workers.

3. General Directorate of Plant Protection and Quarantine (Ministry of Agriculture): This directorate has four functions: research, development, extension, and commercial activities. For its research work it divides the country into six regions (Ankara, Adana, Diyarbekir, Istanbul, Izmir, and Samsun), each with a plant protection institute. Research projects receive final approval from the National Plant Protection Research Council. The main research institute has an average staff of about 26 professionals, and a total of about 180 professionals are engaged in research and development work. For its service, extension, and control functions the directorate has 10 regions covering 49 provinces, each with a director of plant protection and quarantine. The activities of commercial companies in this field are increasing, and 84 commercial aircraft companies are currently in operation.

4. Cotton Directorate (Ministry of Agriculture): This directorate has research, development, inspection, and extension functions. There are regional cotton inspectorates in Izmir, Antalya, and Adana working in

collaboration with the General Directorate of Agriculture. Research is concentrated mainly on problems of breeding, cultural practices, and plant nutrition. There are two main centers of research--the regional cotton research institutes of Adana and Wazilli. Both are associated with major irrigation programs.

5. General Directorate of Soil Conservation and Farm Irrigation--Toprak Su (Ministry of Village Affairs): Toprak Su, which in 1963 was transferred from the Ministry of Agriculture to the newly organized Ministry of Village Affairs, consists of a control organization of five divisions, plus 16 regional offices, experimental and research stations, and soil laboratories. In addition there are branch offices in 15 of the country's 67 provinces. Toprak Su employs about 670 agricultural graduates. Most of the work is concentrated on problems of irrigation and drainage, and relatively little work is done on soil conservation. It is largely on-farm work financed mainly (about 80 percent) through farmer credit from the Agricultural Bank. Training schools for Toprak Su's staff are operated in Tarsus, Izmir, and Ankara. There is a machinery and equipment center at Ankara. Research is carried out by Toprak Su at two soil and water research stations, paralleling similar research by the Ministry of Agriculture. Toprak Su plans three new regional stations with substations. Plans are proceeding to expand research in all of Toprak Su's geographic areas of activity, which will further deplete the supply of qualified research scientists in Turkey.

6. Sugar Factories Company (Ministry of Commerce): The Sugar Institute, under the supervision of a general director, is the research arm of this company. The institute, which employs about 40 university graduates, is divided into two main branches--agricultural research and technological research. The agricultural research branch covers soils and fertilizers, agronomy, plant breeding and entomology, and machinery. It also provides in-service training for the field and factory staff. An associated organization, the Seed Breeding and Improvement Corporation, has a central laboratory and three experimental stations. The agricultural development work is carried out by 126 agricultural district chiefs scattered among the 200,000 farmers contracted to grow sugar beet. The farmers are supplied with farm requisites as well as cash credit for labor. The advances are deducted by the factories from the price paid for sugar beets.

7. General Directorate of State Farms (Ministry of Agriculture): This directorate operates 22 farms in 16 provinces with a total area of about 340,000 hectares. About 270 university graduates are employed, including 235 agriculturists and 12 veterinarians. Seed multiplication is the main focus of the work. About 100,000 hectares are used for this purpose annually, producing about 120,000 tons of seed. An equal amount is produced through contracts with large farmers. It is expected that seed multiplication will more and more be left to commercial interests and the large farmers. The state farms could then be used for other purposes, such as research and demonstrations. Animal breeding is the second important activity of the state farms. This activity is comparable to that of the Veterinary Directorate's stud farms.

8. General Directorate of State Meteorological Activities (Ministry of Agriculture): This directorate runs the State Meteorological Service, which has some important agricultural duties. It employs 33 agricultural technicians and 21 agricultural school graduates. There are plans to produce more detailed climatological data and improve the quality of its agrometeorological services.

9. Ministry of Forestry: This ministry is new, and will have two general directorates--Forest Production and Industry, and Afforestation and Erosion Control. All forests in Turkey are owned and operated by the state. In 1969 the ministry employed about 1,800 graduate foresters and 1,300 forest technicians. Of the graduates about 70 are employed in research. There are two government research institutes, the Forest Research Institute, in Ankara, with four regional stations, and the Poplar Institute, in Izmir, with nine regional stations. In addition the forestry faculty of the University of Istanbul carries out research.

10. General Directorate of Agricultural Equipment and Supplies (Ministry of Agriculture): This agency is a state economic enterprise dealing with the supply of machinery, equipment, fertilizers, plant protection chemicals, and vaccines.

11. General Directorate of the Feed Industry (Ministry of Agriculture): This agency, established in 1958, has 10 feed mills (4 with government ownership and 6 with government-private ownership), producing feed for cattle, sheep, and poultry. Each mill has a small number of extension agents advising farmers on the use of feed. In 1965 the total output of the mills was 60,000 tons, or about 1 percent of the total quantity of concentrates

consumed by animals in Turkey. The corporation purchases raw materials in large quantities when prices are reasonably low, and can sell a product of good quality at a reasonable price without being subsidized.

12. General Directorate of the Milk Industry (Ministry of Agriculture): This directorate, established in 1964, has set up four dairy plants, in Adana, Izmir, Istanbul, and Kars. Switzerland has given the directorate a long-term, low-interest loan for machinery and a grant for technical aid. The milk plant in Ankara is owned by the Ataturk Farm. According to future plans, all milk plants are to be private enterprises. The World Bank-supported project for intensive dairy production in the Ankara, Adana, Izmir, and Istanbul areas will affect the dairy plants by increasing the milk supply to the main consumption centers.

13. General Directorate of Wool and Mohair (Ministry of Agriculture): This directorate has administrative, regulatory, and marketing functions. It was established primarily to promote self-sufficiency in fine-staple textile wools. Premium prices are offered for merino-type wool produced from crosses between merino rams and native sheep.

14. Livestock Development Program Organization (Ministry of Agriculture): This agency, recently set up, is headed by a livestock development program director and has an interdepartmental executive committee. There are separate managers for each project. In the case of the intensive dairy project (supported by the World Bank) there will be four subprojects, one for each project area (Adana, Ankara, Izmir, and Istanbul), with field staff carrying out essentially extension functions. For the eastern part of the country a village livestock development project is planned which also includes the improvement of mountain pastures. This project is expected to cover the three provinces of Kars, Erzurum, and Ohri.

15. Soil Products Office (Ministry of Commerce): This agency (TMO) acts as a national grain corporation. For its major grain operations, purchasing, storing, selling, transporting, and price stabilization, it is organized into nine regional branch organizations. The agency employs about 80 agricultural graduates and about 140 technicians.

16. Meat and Fish Organization (Ministry of Commerce): This agency, established in 1953, is concerned with the production, procurement, and distribution of livestock and fish products, and with some livestock

improvement activities. It was established mainly to encourage fat-stock breeding and to regulate the meat market. The organization has its own slaughterhouses, cold storage plants, and marketing facilities. Credit is extended to producers in order to encourage the fattening of cattle. Market regulation is designed to prevent excessive seasonal price movement. The organization has an ambitious program for expansion, including plans to control 30 to 50 percent of the country's meat trade. Through its five factories it currently handles 5 to 7 percent of the country's meat trade, and employs about 80 veterinarians and two agricultural graduates for meat inspection and commercial management. Livestock are bought directly through agents or on a contract basis, mainly in the eastern part of the country. The organization also exports sheep (350,000 head were exported in 1967 to Iran and Lebanon).

17. Agricultural Bank of Turkey (Ministry of Economics and Commerce): As noted previously, this bank provides the major portion of institutional credit available to farmers, either directly or through cooperatives. Recently an "Encouragement and Development Loans Program" was established which is expected to increase loans to agriculture significantly. The loans will be directed at specific projects, some of which will be supported by IBRD. The bank employs about 70 agricultural graduates and 80 agricultural technicians, mostly in the provincial offices. The bank also administers the law concerning the establishment and operation of credit cooperatives.

18. Planning and Economic Research Department (Ministry of Agriculture): This organization was established in 1963. Its purpose is to organize the efforts of the Ministry of Agriculture in relation to the five-year development plans and also to conduct research in agricultural economics. It is an organization that is directly responsible to the minister. A regulation has been prepared outlining the principal responsibilities, duties, and authorities of the organization. The organization is mainly responsible for the following: organizing the planning activities throughout the Ministry of Agriculture in accordance with the aims and principles of the development plan; developing the targets of the agricultural sector for inclusion in the general plan; accomplishing follow-up evaluations of the implementation of the planned activities; performing economic research in all aspects of agriculture; collecting and evaluating agricultural statistics; and coordinating and implementing community

development activities. In order to carry out these duties the organization has been established with five sections: plan, program, and project preparation and evaluation; plan and program implementation and coordination; economic research; agricultural statistics; and regional and community development. Although it has been possible both through the U.S.-AID program and other technical assistance programs to send some of the technical personnel abroad for training and experience in various fields, assistance has been insufficient for the rapidly growing needs in Turkey. The present staff numbers about 30, of which two have a Ph.D.

19. General Directorate of State Hydraulic Works (Ministry of Energy and Natural Resources): This agency (DSI) is responsible for the construction of dams and large-scale irrigation development projects.

20. National Productivity Center (Ministry of Finance): This center has both an industrial and an agricultural section. The latter tries to coordinate and stimulate efforts to increase agricultural productivity through seminars. It employs 16 agricultural graduates.

WHEAT RESEARCH AND TRAINING PROGRAM

Turkey has a countrywide wheat program aimed at full self-sufficiency in wheat production, taking into consideration the high rate of population increase. The program consists of three projects: wheat research and training; seed production and distribution; and increased wheat production.

The wheat research and training project is supported by the Rockefeller Foundation on the basis of an agreement signed in 1969. The foundation provides four foreign staff members, with Dr. Bill C. Wright serving as the project's leader. In addition there may be temporary staff added from time to time for specialized assignments. The project, which was established to increase both wheat yields and the profitability of the production of winter- and spring-type bread wheats and durum wheats, works in the closest possible coordination with the International Maize and Wheat Improvement Center (CIMMYT).

In order to accomplish its goals the project has the following specific objectives: to organize a smoothly functioning, nationally coordinated wheat improvement program; to develop new varieties and cultural practices; to develop a capability in the surveillance and control of plant diseases and pests; to develop the scientific basis and techniques for a program of soil and moisture conservation and crop management for each ecological area of Turkey; and to organize an in-service training program designed to improve the scientific capability of staff engaged in wheat improvement and production programs.

The project provides for 11 research stations in the major wheat-growing areas of Turkey. There are three principal breeding stations--one in Ankara and one in Eskisehir, for winter wheat, and one in Izmir for spring wheat. The eight remaining stations are largely selecting and testing sites. In addition, off-station tests are performed on farmers' fields and off-season breeding nurseries are operated for spring wheat.

In Turkey there are two distinct wheat-growing areas: the dry, cold winter-wheat areas of the central plateau and southeastern Turkey; and the warmer, wetter spring-wheat areas of the Turkish coasts. The production problems in the two areas are entirely different. In the winter-wheat area there are a number of improved varieties which are yielding less than their potential, mainly because of inadequate rainfall and the poor management applied to the fallow period. The most pressing problem in the winter-wheat area is to develop a set of tillage practices that will permit both the establishment of early wheat seedling (preferably germinating from residual moisture) and the maximum conservation of rain falling on the fallow land.

In the spring-wheat area the most pressing problem is to develop varieties that have good resistance to such diseases as stripe rust (yellow rust) and septoria, and durable varieties that can be sown in early October, and yet will flower in April.

From a wheat breeder's viewpoint it is necessary to distinguish between nine ecological areas for wheat growing in Turkey (see Figure 5). There are some significant differences also among the various coastal areas. In Thrace only winter wheat is grown.

Table 8.2 shows the distribution of the wheat area in Turkey by regions and by types of wheat.

In 1963 many of the coastal areas grew a relatively large amount of durum wheat. The later introduction of Mexican wheats mainly replaced the durums. The price of durum wheat is 9-10 percent higher than the price for bread wheat.

The breeders have been making a determined effort to incorporate as much diversity into the breeding material as possible by acquiring exotic germ plasm from around the world and carrying on a wide crossing program. In the spring-wheat breeding program, centered in Izmir, a large number of lines from many different parts of the world have been screened for their resistance to septoria and stripe rust as well as their yield potential.

TABLE 8.2

Wheat Area in Turkey

Region of Turkey	Percentage of Total Wheat Acreage	Percentage of Acreage in Each Type: Durum Wheat	Bread Wheat	Compactum Wheat
Thrace	3.7	60	38	2
Black Sea	2.7	n.s.	100	n.s.
Marmara South	5.0	25	75	n.s.
Ege	2.5	75	25	n.s.
Mediterranean	4.9	40	60	n.s.
Central plateau	46.0	30	30	40
West transitional	5.4	60	30	10
North transitional	7.0	65	30	5
East transitional	6.7	30	50	20
Southeast	10.4	60	40	n.s.
East	5.7	n.s.	80*	20*
All areas	100.0	37	41	22

*60 percent planted in spring.

Notes: Percentages by regions are for 1963; percentages by types of wheat within regions are as of April 1968. n.s. = Not significant amounts.

Source: Percentages by regions calculated from data supplied by the Ministry of Agriculture. Distribution of types within regions based on estimates of committee working with Rockefeller Foundation group, April 1968.

Also a series of tillage experiments are being conducted consisting of six experiments at four locations. The objectives of this series of experiments are to develop improved tillage practices for managing the fallow so that the maximum amount of moisture is conserved, and to study fertilizer use and other management practices. This tillage research is a joint effort by Oregon State University, AID, and local Turkish organizations.

FIGURE 5

Nine Ecological Regions for Wheat Growing in Turkey

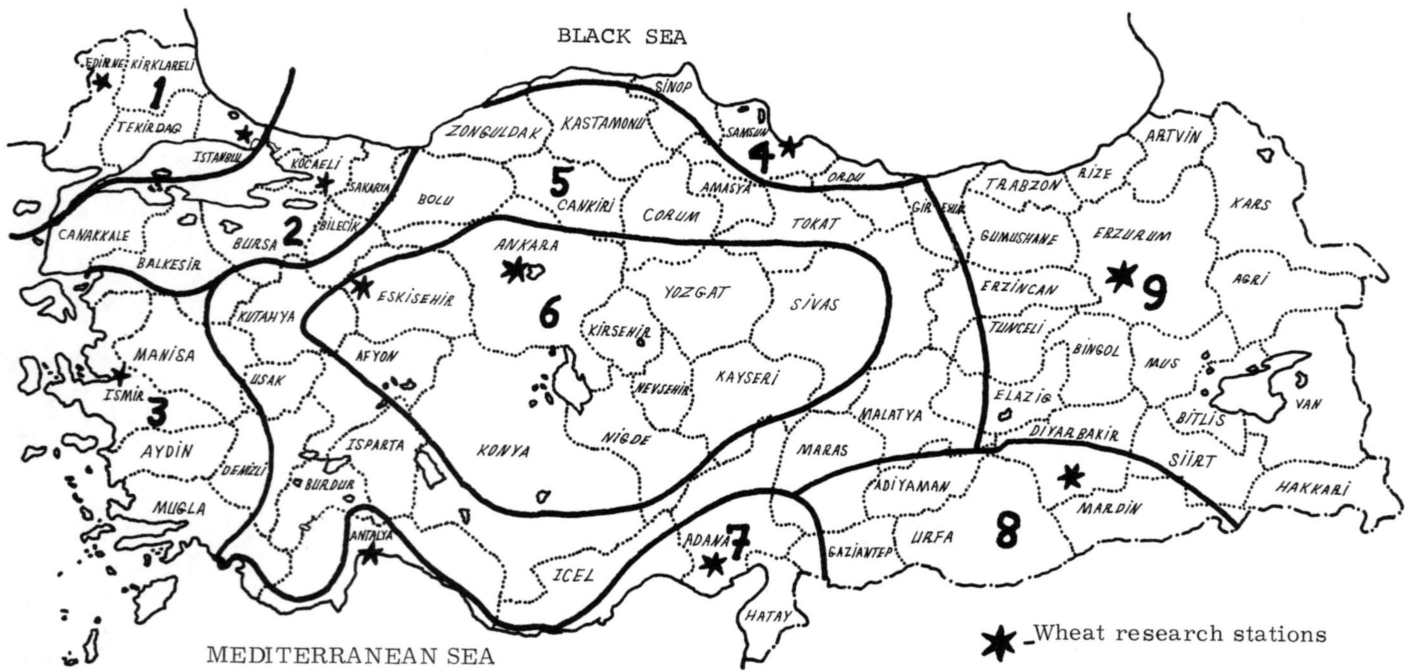

Type of Wheat	Date of Planting	Date of Heading	Date of Ripening	Region
Winter-durum & bread	October-November	Mid-May	Early July	Thrace
Spring-bread	November	Late April	Late June	Marmara
Spring-bread & durum	Mid-November	Mid-April	Mid-June	Ege
Spring-bread	November	Late April	Late June	Black Sea
Winter-bread & durum	October	Late May	Late July	Transitional
Winter-bread & durum	15 October-15 November	Mid-May	Early July	Central Plateau
Spring-bread	Mid-November	Mid-March	Late May	Mediterranean
Winter-durum & bread	November	Mid-April	Mid-June	South East
Spring-bread planted spring	Late April	Early June	Mid- to late August	East

Source: The Wheat Research and Training Project, Turkey.

During the 1970-71 wheat season an annual mobile disease survey and a series of disease-trap nurseries were established. In the 1971-72 season the disease survey was greatly expanded with the full cooperation and assistance of the General Directorate of Plant Protection and Quarantine.

Training is a very important part of any wheat program, and in Turkey a comprehensive training program has been planned that will be carried out in three steps. Initially the project involves the training of national- and provincial-level wheat specialists. These specialists will then train county agents and village technicians for local work; the local agents will in turn carry on the farmer training program.

The training of the national- and provincial-level wheat specialists will consist of classroom and field work, and will be carried out by the Wheat Research and Training Center near Ankara with the help of the AID team. The field program will take place at the state farms and research farms. It will provide training and practice in each step involved in the establishment of wheat production demonstration plots, with emphasis on moisture conservation practices, improved seed bed preparation, the use of high-yield varieties, optimum fertilizer use, and weed control. This program will be repeated each year to train additional provincial-level workers.

The entire program, but especially the portion involving farmer training, will make use of radio, television, newspapers, and other communication and extension possibilities. In addition to the establishment of demonstration plots, adaptive research activities will be carried out on the farmers' fields by the trained provincial specialists under the control of experts from the wheat research and training project. The results of the research will be transferred to the farmers immediately: the strategy of this program is to teach wheat specialists, county agents, and farmers improved practices as they are developed.

CHAPTER

9

THE EDUCATIONAL INSTITUTIONS

The manpower employed in agriculture and related activities in Turkey can, in broad terms, be grouped in the following three basic categories: technically trained staff employed in the public or quasipublic sector; technically trained personnel working in the private sector; and the majority of the farmers, farm family members, and hired farm workers, with no formal training in agriculture. Since the educational level of the third group is relatively low, with a high percentage of illiteracy, both the quantity and the efficient use of the technically trained personnel are of crucial importance for agricultural development.

The educational institutions responsible for the development of agricultural manpower in Turkey offer a variety of programs, including the following:

1. Agriculture--university level: Three university faculties of agriculture are now functioning in Turkey. They are all relatively new. The former Institute of Agriculture in Ankara, which was under the Ministry of Agriculture, provided a practical training program that dated back to 1933. It may be considered the progenitor of the present system, since the different departments of this institute became the current faculties of agriculture. The faculty of agriculture at Ankara University is the oldest of the three faculties, dating back to 1945. The faculties at Ege University in Izmir and at Ataturk University in Erzurum are still in the process of development, having been established in 1965 and 1958, respectively. A fourth faculty of agriculture is being planned as the initial faculty of the multifaculty

University of Adana. The necessary legislation for this was passed in 1969, and a dean and 11 key staff members have been appointed from the Ankara faculty. A fifth faculty of agriculture is being planned for establishment in 1975 at Diyarbekir in southeastern Turkey.

The total capacity of the three operating faculties is approaching 3,000 students, with the Ankara faculty enrolling about half of these students. This faculty seems to be overcrowded: the average number of students graduating in the last five years has been about 420 per year.

Since 1965 the three operating faculties have changed from a four-year study program to a five-year program, with specialization coming in the last year or two years, depending on the faculty. This emphasis on specialization and a system of elective courses is seen as the result of the influence of U.S. educational experts. The Nebraska AID team has been assisting both the Ankara and Ataturk faculties since 1955.

The three established faculties are all provided with, or have plans approved for, adequate living accommodations, buildings and land for research, and facilities for demonstration and practical student training for a half-year or longer.

The organization of the agricultural faculties at Ankara and Izmir is mainly of the "traditional" type, and includes a number of relatively small "laboratories." The Ataturk faculty is organized on a department basis, which is better suited for specialization and for graduate work. Data concerning the staffing of the different faculties are given in Table 9.1.

The academic selection system effectively discriminates against students from the educationally less-favored rural areas. Also the best students, even those of rural origin, are rarely guided to agricultural faculties. As a result only 10 percent of the agricultural students are of rural origin. This works against their effectiveness as future field workers in remote rural areas.

The faculty of agriculture at Ankara University established a postgraduate program in 1968. The program is organized within the framework of the 26 departments of the faculty.

Hacettepe University, Ankara, which has no regular agricultural faculty, is in the process of establishing an exclusively postgraduate faculty of agriculture with the following aims: to provide students with high

TABLE 9.1

Number of Educators in Faculties of Agriculture, Veterinary Medicine, and Forestry in Turkey, November 1969

Faculty and Year of Establishment	Professors	Assistant Professors	Doctoral Assistants (Lecturers)	Other Assistants	All Educators
Agriculture					
Ankara University (1945)	41	38	40	63	182
Ege University, Izmir (1965)	16	15	29	40	100
Ataturk University, Erzurum (1958)	9	10	10	24*	53*
Adana University (1969)	6	6	0	0	12
Diyarbekir University (1975)	0	0	0	0	0
Hacettepe University (1969)	0	10	0	0	10
Total	72	79	79	127*	357*
Veterinary Medicine					
Ankara University	33	26	14	24	97
Istanbul University	0	0	0	0	0
Elazig University	0	0	0	0	0
Forestry					
Istanbul University	22	14	11	14	61
Karadeniz Technical University	0	0	0	0	0

*Plus 56 part-time assistants.

Source: State Planning Organization, Turkey.

professional qualifications, and to train scholars for academic positions. This postgraduate faculty will ultimately include 15 departments or institutes with a full-time staff of 75. The total number of students will be about 75. The dean, Dr. Osman Okyar, was appointed in January 1970. Applied research may be carried out at already established research institutes, under the guidance of the Ministry of Agriculture, or at the research farms of other universities. The program will place emphasis upon competence in a foreign language and the basic sciences, which will serve as a foundation for scholarly work in higher applied studies and research.

2. Agriculture--technical and vocational level: Technical and vocational education in agriculture is going through a transitional period of upgrading and expansion. Until 1963 this type of education included a lower secondary level (grades six to eight), at eight technical schools, and a higher secondary level (grades nine to eleven), at four regional agricultural schools.

Graduates of the regional agricultural schools were required to enter the service of the Ministry of Agriculture for a minimum period of five years, following the completion of their studies. Later all technical agricultural (and horticultural) schools were changed into a single type--at the higher secondary level (grades nine to eleven). At the end of 1969, 34 such schools were operating with a total capacity of 7,550 students. The aim of these schools is to provide one such school for each province. The number of applicants is quite high, about eight to nine times the number of openings at the previous schools. The capacity of each school ranges from 150 to 300 students (mostly in residence).

In theory the schools serve the dual function of training technicians for public service, and training potential farmers. In practice nearly all the graduates seek employment in the public sector, to which they have a legal right. The idea that the school provides useful training for potential farmers does not seem to have gained wide acceptance.

For nearly all students in these schools education is totally free, including their board, lodgings, and school uniforms. The teacher-student ratio seems to allow for a great expansion of enrollment without the further addition of staff. Recently many schools have been established on the state farms and in research institutions. This has reduced significantly the needed capital investment. Often the director of the institution is also the director of the school.

The General Directorate of Soil Conservation and Farm Irrigation (Toprak Su) has three schools for the in-service training of its staff and for the training of farmers as farm extension workers in the new irrigated areas. These farm extension workers are given six weeks of training.

3. Forestry--university level: The forestry faculty was integrated into the University of Istanbul in 1948 (previously it was a department of the Institute of Agriculture, in Ankara, but located at Istanbul), and now it has a total of about 1,100 students. It graduates some 125 students annually, offering a four-year program. The total number of graduate foresters in Turkey is now about 2,050, and the faculty plans to reduce its intake somewhat. The first year of study is undertaken in the science faculty at the University of Istanbul.

Karadeniz Technical University in Trebzon (established in 1965) has a faculty of soil sciences, in which there is a small department of forestry that plans to accept students for a five-year period.

Since all forests and most forest industries are state owned, employment opportunities are mainly in the public sector and primarily in the Ministry of Forestry. At present no formalized postgraduate program is planned in forestry.

4. Forestry--technician training: There are currently two forestry schools providing a four-year course (grades nine to twelve). These schools are located at Duzce and Trabzon, and were founded in 1958 and 1964, respectively. The capacity of both schools is 200 students and recently admission has been restricted. The number of applications for admission have been unusually large, about 1,000 applicants for 50 openings. About half of the students are said to have a village background. The education is completely free, but in return the students are bonded to the forestry service for eight years after the completion of their education. To date the total namber of graduates from the two schools is approximately 500.

5. Forestry--vocational training: There are at present four schools providing one year of training for students at the primary level who have had adequate forestry experience before entering the vocational schools. The total annual capacity of these schools is about 230 students. The training has a strictly practical orientation.

6. Veterinary science--university level: The only veterinary science faculty has been attached to Ankara

University since 1947 (previously it was part of the Agricultural Institute in Ankara). It offers a five-year course. Technical agriculture and animal husbandry figure strongly in the program. This faculty currently enrolls about 400 students, and there were 62 graduates in 1969. Practically all graduates go into government service. It is reported that at present only 15 veterinarians are privately employed.

A rapid growth in the demand for veterinarians is foreseen in the Second Five-Year Plan. By 1985 it is estimated that the total requirement for veterinarians will be about 3,000. This is based upon a ratio of one veterinarian for each 12,000 livestock unit equivalents.

In order to meet these requirements a rapid expansion of the training capacity has been planned. In Istanbul University a second veterinary faculty began operating in the 1970/71 academic year. A third faculty was started a year later at Elazig University. Each of these faculties has a planned annual enrollment of 100 students. Two additional veterinary faculties have been proposed.

At present no formalized postgraduate veterinary education is provided, nor are there postgraduate degrees awarded. The consideration of a postgraduate veterinary school, which could be established in conjunction with one of the institutes of Hacettepe University, would seem to deserve high priority.

7. Training of animal health officers: There are at present three training schools for animal health officers, located in Selimiye, Samsun, and Konoya. They provide a three-year training course (grades nine to eleven). Each school has an annual admission of 50 students, most of them reportedly with a village background.

While this education is free graduates are bonded to the Veterinary Service for five years on the completion of their course. At present the total number of animal health officers employed in government service is about 1,100. A total of about 7,000 may be required by 1985 to serve along with the 3,000 veterinarians that are targeted for that year.

8. Home economics training: There are no training facilities in home economics at the university level or equivalent levels. About 20 women graduates in agriculture have specialized in home economics and are working in the Ministry of Agriculture. There are five home economics schools at the upper secondary level with three-year programs. The total capacity is about 700 with a potential output of about 230 per annum. These schools

provide the staff for the Technical Agricultural Organization. At present about 220 are in government service. The indicated target for 1985 is one graduate home economist and two technical-level home economists per district.

9. Fishery training schools: A fishery training school (secondary level) is being established in Istanbul with bilateral Japanese assistance.

CHAPTER

10

THE RURAL COMMUNITY IN RELATION TO PROBLEMS OF DEVELOPMENT

AGRICULTURAL VILLAGES

This chapter is based mainly on literature about and on the author's own observations of the agricultural villages in semiarid central Anatolia: This part of Turkey poses particularly difficult farm problems. However, many of the observations and conclusions in this chapter seem to be valid for a much wider part of Turkey.

One typically finds on the Anatolian plateau a rather tightly knit village community, within which all types of farming take place. This situation seems to have important policy and planning implications, in regard to the introduction of significant changes in the farming system, which have not yet been fully recognized. According to the Second Five-Year Plan, the village can be defined as follows: "a community where the economic structure is preponderantly dependent on agriculture, where social relations peculiar to it exist, as a place with a definite boundary for settlement, the smallest administrative unit and a locality whose population is under two thousand."

There are in Turkey about 35,000 villages plus about 30,000 other places of settlement that are called farms, neighborhoods, outposts, nomad camping sites, etc. About 55 percent of these villages have a population of 151 to 500. Only about 10 percent of the villages have a population of more than 1,000. Of the total number of villages about 72 percent are compact units. About 13,000 villages (37 percent of the total number of villages) with a total population of about 7 million (nearly one-third of the total rural population) are located in forested areas or on the edge of forests. These villages are not easily accessible and are scattered over a wide area.

The living standard in these areas is low and the potential for development is limited. Government efforts to develop these areas have been under the control of the Ministry of Village Affairs, and may now be moved to the Ministry of Forestry, in spite of the fact that the villagers' main income usually comes from agriculture.

Typical Village in Central Anatolia

According to Nicholas Helburn, the "typical village" in central Anatolia is a group of some 80 houses built of stone, around a common water supply.* Every house has at least two rooms, a kitchen and a stable. About 25 percent of the houses have, in addition, a living room. Outside the village there are few structures. There are no fences, no sheds, no barns. Everything is stored in the house or in piles outside. Straw piles are usually covered with six inches of soil.

Each village has common shepherds and cowherds and a common watchman. Usually the shepherds and cowherds bring the flocks and herds back into the village at night, and call for them again before sunrise. The ewes and cows, in the lactation period, are milked at night and in the morning in the owner's stables--by the women of the household. The representative household may have one cow, a pair of oxen, twelve sheep, six or eight goats, one donkey, and some poultry. There does not seem to be any control of the male animals or of breeding practices. The type of selection of bulls that is being practiced seems to have a negative effect, since the large, fast-growing bulls are castrated to serve as draft animals while the smaller bulls are allowed to run with the herd.

All the village land is grazed except the vineyards and vegetable gardens, which are protected by stone walls. Natural pasture held in common by the whole village accounts for about 60 percent of the grazing land. The remaining grazing area is accounted for by small plots of crop land. These plots are individually owned, but are regarded as common grazing land throughout the year when they lie fallow and after the harvest when they are cropped. There is no control of the number of stock that

---

*Nicholas Helburn, "A Stereotype of Agriculture in Semi-Arid Turkey," Geographical Review, vol. 45, no. 3 (1955): 375-84.

any one villager puts into the herd; this contributes to severe overgrazing.

A two-year wheat-fallow system of farming is maintained in almost all of the Anatolian villages. One-half of the village wheat land is sown during a particular year and the other half is left fallow and used for pasture. This system prevents any individual farmer from cropping the same fields in successive years. Each man farms his land independently and he may miss one or more turns in the cycle. He must harvest a field during the same period that his neighbors harvest in order to remove the crop before the flocks and herds are turned loose on the land. This system ties all the villagers together. The introduction of a new crop cycle is virtually impossible for the individual farmer. It can only be done for the village as a whole, because of the common right of all farmers to pasture flocks and herds on the cultivated land only every other year.

There is a close relationship between the area of cultivated land and the number of animals belonging to each farm family. On the average one man and one pair of oxen can plow eight to ten acres a year. The number of sheep and cows is also roughly tied to the scale of household farm operations, by the need to feed them during the winter. Proper feeding would require stall feeding for four to six months. Because of the lack of feed, some farmers cut down the time of stall feeding to two months. The serious shortage of feed is the main reason why the state farms are having difficulty selling their large purebreed bulls to the villages--nobody wants to feed the bulls.

In the central part of the plateau there do not seem to be any nomadic sheep owners. The natural grazing land is completely under the control of the village, with no grazing rights granted to outsiders. The farm families have a relatively common economic interest--improving their income in proportion to their share of cultivated land. There is usually no conflict of interests regarding the joint planning of improvements. In those parts of the country where migration by pastoral nomads takes place, conflicting interests may occur.

The work year in the villages starts in the fall after the first rains, in late September or in early October. The optimum planting time may last for only about two weeks, but the seeding and plowing may go on for a much longer period, in spite of the lower yields from the later plantings. During the winter months weather conditions prevent the farmers from working outside. From the end of November to the end of March they have nothing to do except care for the animals, which are kept in the stables.

The grazing period begins between April 15 and April 30 and continues for about six months, until October 15-30. Farmers who do not have enough winter feed for their animals lengthen this period to 10 months.

When spring comes the farmers' work is devoted to plowing and planting. Planting of spring wheat and barley is followed by the plowing of the year's fallow land. Vineyards, potatoes, and vegetables are planted, and grass growing in odd places is cut, mainly by the women for use as hay. In late May and in June the men are comparatively idle, prior to the start of the harvest in July. The harvest period lasts for about two months. This period is a busy time for everyone. In September the pressure eases until the first rains, when the cycle begins again. The busy seasons for the male farm population add up to about four months each year. This is the explanation for the huge estimated underemployment in agriculture.

The women play a major role in village life. The village woman is not only housewife and mother, but she also works in the fields. Between seasons the men may go to the cities or the towns to seek employment, while the women remain near their fields and animals throughout the year. In many provinces the women workers form the majority engaged in agricultural activities. In the Black Sea region women account for about two-thirds of the agricultural labor force. The fact that the burden of agricultural activities falls mainly on the shoulders of women is related both to the types and techniques of Turkish agriculture and to social factors and female-male ratios. There has been a continuing migration of men to the cities and to other countries. The resulting dominance of women in Turkish farming may have important implications in the preparation of strategy for agricultural development. The rate of illiteracy is even higher among Turkish village women than among men. The women may also be more conservative than the men. Thus the question that may be asked is: what role will women in the villages play in the decision-making process? No attempt has been made to reach these women through special extension programs. Furthermore the author has encountered no studies or discussions of their proper role in the strategy of rural development.

Laws in the Villages

It is a peculiar fact that in the villages three sets of laws are recognized: the Turkish Civil Code, the Seriat (Koranic Law), and local customs. Of the three, the Turkish Civil Code is followed the least, though this is the only code that can be legally enforced. Under the Civil Code legal rights to land depend on a cadastral register. In Turkey, as mentioned previously, there is as yet no cadastre for the main part of the country, and though registration deeds exist, in rural areas they are seldom up to date or complete.

Legal title to land can be established in three ways: through formal title deeds; through tax receipts showing a man to be the owner of the land; and through use of the land for 20 years. What matters most to a villager seems to be whether he can in practice cultivate a certain piece of land and gain acceptance of his customary rights to it from the local community.

The villager can plow up uncultivated common pasture land, and then establish private ownership rights to it after using the land for 20 years, so long as the rest of the villagers allow him to do so. During the first three decades of the Republic, from 1923 to 1953, when land was plentiful, as the sons of a farm household reached maturity the household was able to expand its landholding. However, the supply of cultivable land has been drying up. This new situation has led to an increase in the frequency of sharecropping and in the proportion of landless labor. It also has increased the exodus of rural people to the large cities.

When land changes hands at the death of the owner the settlement of the inheritance seldom corresponds to the Civil Code. Purchasers and heirs do not usually go to lawyers to register changes, because, apart from the costly bureaucratic and legal procedure of altering the deeds, there would be too many questions to answer. The present confusion seems to be self-perpetuating so long as village opinion condemns any resort to court action.

In the case of a multiplot inheritance, when there is more than one surviving son, equal portions of each plot are usually given to the sons. While this system has led to widespread fragmentation of land, villagers have favored it. They feel more protected against droughts and other hazards with their land divided into plots that are distributed throughout the village area. It is considered a great risk to have all of one's land on a single soil type. Daughters' rights to land are regarded as unimportant and are not allowed to prevent a man from possessing enough land to care for his family.

The villagers are in no particular hurry to register new holdings, such as plowed-up pasture land, because the land would then incur tax. In most cases legal title would also be difficult to establish. This legal situation was not of great significance in a subsistence economy, but more recently in Turkey, with the expansion of the market economy, based largely upon farm credit, the unfavorable aspects of the situation have come more and more to be known and felt by farmers.

Widespread Illiteracy

In spite of the government's educational efforts there are still thousands of villages without functioning schools; illiteracy remains widespread. A 1969 study by the Turkish State Planning Organization indicates that only 39 percent of the Turkish villagers above 16 years of age claim literacy (the ability to read and write their own names). Only 10 percent of the villagers over 16 claim to have graduated from primary school (the fifth grade). Children finishing the village elementary schools have a very limited opportunity to attend institutions of a higher level. Lack of reading material in the villages has made it difficult to maintain literacy. Reading rooms have now been opened in more than 6,000 villages, and more than 4 million books have been distributed to other villages. Almost half of the total number of villages receive newspapers at least once a week. In many other countries the mass media have been effectively used to accelerate modernization and agricultural development, but in Turkey there seems to be great variations in regional access to mass communication, with the Marmara region showing the most penetration by the media and the southeastern region showing the least penetration. About 3 million radio sets are in operation in the villages.

Telephone facilities in the villages are very scarce; in 1963 less than 5,000 villages had any telephone facilities whatsoever. The postal service is also very limited. In 1967 the villages had only 24 main post offices, 140 post office branches, and 136 post office agents. In addition postal services were rendered through about 2,000 mail carriers.

The villagers' opportunities to benefit from medical treatment have remained limited. At the end of 1966 health services were provided to about 10,000 villages located in 17 provinces. Only 50-60 percent of the village dwellings have glass windows, and the great majority of dwellings lack the simplest facilities for taking a bath. The number of public baths and public toilets is very limited. Fuel is a major problem, with animal dung being the main source. There is usually a mosque in every village.

Drinking water is still a problem in many villages. According to the Second Five-Year Plan, 25,000 of the 65,000 rural settlement units, or 30 percent, lacked drinking water, and 19,000 units had inadequate drinking water in 1966.

Appropriations from public funds to provide electricity for villages were made for the first time in 1964. In the beginning of 1967 only about 700 villages were equipped with electric power.

The system of village roads is insufficient. In 1966 the total length of the standard village road was about 23,000 kilometers, far short of the required 150,000 kilometers.

Sociological Implications

The foregoing description of the villages suggests some important sociological implications for the work of the agricultural field services. The typical situation in the villages calls for a special strategy in the introduction of new technology. One has to work with the village as a unit, not with the individual farms. In tightly knit village communities this may not be too difficult, assuming that a proper strategy is used--this does not seem to have always been the case.

It may even be misleading to use the word "farm" (ciftlik) for the Turkish situation. There are virtually no farms in Turkey in the sense of a compact and permanent unit of agricultural production. Instead the typical unit of agricultural production, the "operational holding," is an aggregate of generally scattered parcels of land. These operational holdings must not be thought of as permanent farms, but rather as temporary aggregates of land and as rights to land. The number, size, and ownership of parcels of land, and their grouping into operational holdings, are in a constant state of flux. The only stable and permanent unit to work with is the village. Nearly every village has an official document in which the main points and lines of its boundaries are noted. However, in some cases the village's total land area has not been measured.

If, for example, a promising new system of wheat-legume rotations is developed, it has to be introduced through a series of demonstrations performed for a "model" village. The farmers of such a village would first have to be brought to research farms and state farms to observe this new technology and to be convinced of its merits. In addition they would have to be provided with incentives to participate in a demonstration program involving the entire village. These incentives might be in the form of free seeds and fertilizers, or easy credit. If large profitable differences were demonstrated through the use of the new crop rotation, other villages might adopt the system with less incentive than that given to the model village. Thus the problem is one of working with an entire village community rather than with individual farmers or farm families. The introduction of new cropping patterns in Turkey will require proper planning and strategy consistent with the actual village situation; if this can be done, a model would be created for other countries with village communities of about the same type.

To date, however, the attempts to introduce change in Turkish agriculture do not seem to have been highly successful. It has been reported, for example, that demonstrations of improved pasturing by the Technical Organization of the Ministry of Agriculture have come to naught because fences were pulled down by the villagers, resulting in undisturbed overgrazing. It is likely the villagers regarded their action as a legal protection of their grazing rights. Possibly the failure of this effort could have been avoided by working through the muhtar (village headman) and the hoca (imam, or prayer leader, of a mosque) to achieve support for the project from the entire village.

The mounting population in the villages has led, on the average, to smaller farms per family, and to a larger percentage of sharecroppers and landless labor in the villages. An increasing number of men have to leave the villages in the off-season periods to earn an income in the cities, or they settle down in the gecekondu settlements. Since the land resources of each village are fixed, increasing population can only lead to less land per family, increased risk of poverty, and a further impetus to leave the villages. Under the present social system the peasant finds himself at the bottom of the national social scale. This gives him an additional reason to try his luck in a city.

On the other hand an increase in the exodus to the cities may depress the standard of living for the urban newcomers until a relative balance is achieved. Assuming an annual growth rate of 6 to 7 percent in the gross national product, the rate of growth in employment in other sectors may be insufficient to employ all the surplus labor from the rural sector. Through the rapid growth of the cities the village population may, however, hopefully reach a peak in a decade or two, if, in addition to a high economic growth rate, economic and social policies are geared toward a more balanced social system in the main sectors of the economy. The new technology, especially farm mechanization, may lead to imbalances in the villages if necessary social adjustments are not made.

REGIONAL VARIATIONS IN RURAL DEVELOPMENT

A recent study entitled "Regional Variations in Rural Turkey," conducted by the Center for Informational Studies of the Massachusetts Institute of Technology, underlines the sharp differences in rural development among the various regions of Turkey. The variations that prevail in village characteristics from one region to another range from the types of business establishments in a village, such as coffeehouses and retail shops, to the rate of literacy. In the Black Sea region there are many scattered farms instead of the usual clusters of farms found in other regions. The Aegean and Marmara regions are the most developed rural areas, with the south central, Mediterranean, north central, and Black Sea regions at an intermediate level of development, and the northeastern, east central and southeastern regions showing the least development. The impact of relative isolation upon rural development seems to be strong. Proximity to the west and to the coastal location seems to be the main conditioning factor, with the degree of rural development declining from west to east and from the littoral areas to the interior areas. The relative number of agricultural laborers as compared to farm owners or tenants is greatest in the most modern regions, indicating a more capitalistic type of agriculture in these regions.

The size of the farm household and the degree of residential "crowding" reach a peak in the less developed regions. The peasants in the more developed regions are less likely to want large families than those in the more backward regions. Rural modernization may therefore help to reduce population expansion. The MIT study, based upon interviews with about 6,000 peasants, indicates that only 33 percent, on the average, want large families (five or more children). This suggests that population may be controlled not only through rural development, but through the spreading of information on family planning.

The typical Turkish farmer has become a small landowner, growing more cash crops for a larger and wider market, including export markets. The farmer's social relations have become more complex through the influence of mass communication, through greater contact with government officials and with migrant labor, through increased social services in the villages, and through the carrying out of such responsibilities as military service and the power to vote. The farmer represents the majority of the voters and according to the rules of democracy the political power belongs to him. He is, however, more likely to be illiterate than literate. His farming techniques are slowly changing and becoming more dependent upon imports from outside the village, such as new crop varieties, fertilizers, and pesticides. Literacy is becoming more and more important to him as a productive farmer.

PROBLEMS OF DEVELOPMENT RELATED TO CHARACTERISTICS OF THE RURAL COMMUNITY

Large parts of Turkish agriculture retain a traditional character, being comprised of small marginal family enterprises which are partly subsistent, and in which the production effort and the mode of living are almost interwoven. This type of agriculture, which is mainly found along the northern and southern mountain ranges and in the mountainous areas of the east and of southern Anatolia, has continued to exist because of the typical small holdings and poor, overpopulated villages, and because of the long distances to the markets and the lack of transportation and communication.

Modern agriculture has developed initially in the better farming areas, and mainly through the following efforts: the use of new improved technology, such as that represented by the sugar beet plants in central Anatolia, and the fruits and vegetables along many of the valleys; the establishment of tea and tangerine orchards on the coast of the Black Sea, and citrus and banana orchards, and truck gardens on the southern coast; and the introduction of new high-yield varieties of crops such as the Mexican wheats, together with fertilizer, more and improved irrigation, mechanization, plant protection chemicals, and so forth.

The most modern and commercialized agriculture is found in Adana and Anatalya provinces in the Mediterranean region, along the river valleys running toward the sea in the Aegean region, and on the plains south of the Marmara Sea in the Marmara region. Large- and medium-size farms have led the modernization process. The process has mainly progressed for plant production, while animal production has remained more stagnant, with a limited introduction of new technology.

The differences in modernization between various areas and groups of farmers are influenced by the risk factor. The risk in agricultural production in Turkey is much higher in the dry, rainfed areas than in the coastal regions, which have higher rainfall and relatively more irrigated land. Small farmers with a marginal standard of living have been unable to take the economic risks involved in innovations. A shortage of capital and savings has further reduced the ability to invest and take risk. The lack of education is another factor. In 1970 the literacy rate in Turkey was still around 55 percent with the lowest percentage in the more remote rural areas. Compulsory education was still not enforced in 15,000 villages located in the less-developed areas. Approximately one-third of the villages are located in mountain and forest regions with a lack of transportation and communication facilities.

CHAPTER

11

AGRICULTURAL PROJECTIONS

METHODS AND ASSUMPTIONS FOR OECD PROJECTIONS

This chapter is based upon a report entitled "Agricultural Projections for 1975 and 1985," issued by the Organization for Economic Cooperation and Development (OECD); and the "Country Study" for Turkey by the same organization. The "Country Study" was carried out during 1967 and the first half of 1968, and includes information available up to this period. The OECD Secretariat has discussed its projections with various authorities of member nations, and the findings have been considered by the OECD Committee of Agriculture. However, this report and the country studies do not engage the responsibility of the member countries, nor do they necessarily represent their views or the views of the organization. The responsibility is the Secretariat's alone.

Regarding the methods and assumptions adopted by OECD in making its projections, it is important to bear in mind the following points from the report:

1. It has been assumed that present policies will remain broadly unchanged. However, account has been taken of the changes in policies which are already in effect and which can be expected to continue.
2. In assuming the continuation of present policies, the study does not necessarily assume that the objectives of these policies will be achieved. The assumption relates essentially to policy measures.
3. It has also been assumed that trends in prices and costs will continue to develop in approximately the same way as in the recent past. In almost all cases

relative prices have been changing, and the most neutral assumption is that these price movements will continue in the same way. In most countries this approach has led, for example, to projecting lower consumption of beef and higher consumption of poultry meat than would have been the case with a constant price assumption.

4. The aim has been to project the production likely to arise from the continuation of recent trends, under the general assumptions indicated above concerning policies and prices.

5. The difference between projected production and total utilization does not represent a projection of net trade as such. It has been termed "balance" in the tables, and corresponds to potential net export availabilities or net import requirements.

The basis for setting crop yields is that only moderate increases in yields are projected for 1975 and 1985, except for wheat, where new developments already were underway when the projections were made. It is stated in the country report that "on poor land with an erratic rainfall, on which much of Turkey's food is produced, improved varieties and fertilizers are of little avail, assuming the farmer could afford to pay for them." Improved wheat varieties are expected to be introduced only in irrigated and higher rainfall areas (1.0 million ha in 1975 and 1.5 million ha in 1985).

The projections for meat production are based upon the following reasoning:

1. The rising trend in the number of cattle is assumed to continue, together with a slowly increasing percentage of milking cows.

2. A continued increase is also projected in sheep numbers, leveling off slightly because of expected forage problems.

3. An accelerated downward trend is projected for ordinary goats; a slower decline is projected for angora goats.

4. For buffaloes a decline is projected after 1975.

5. The meat production per animal of each type was assumed to remain constant, except for cattle and buffaloes, where a slight improvement in the ratios was assumed.

The projections for milk production are based upon the following reasoning:

1. The yield in the base period was extremely low.

2. Through improved animal husbandry and the release of dairy cattle from draft work, it should be possible to achieve higher milk yields.

3. Higher milk yields per cow than could be derived from recent trends have been projected, but without assuming any rapid increase through the introduction of new technology. The yields therefore would remain extremely low for 1985 (only 722 kg per year for milking cows).

4. Some increase is projected for buffalo cows, but no increase in average milk yields of sheep and goats has been assumed.

5. As a result of the increasing share of cows' milk in total production, the average fat content of all milk produced is expected to decline from 4.75 percent in 1961/63 to 4.52 percent and 4.31 percent in 1975 and 1985, respectively. A similar decrease in average solid non-fat content is projected. The same coefficient for conversion of milk production to butterfat that was used for the base period has been used for the whole period. No increase in processing efficiency is projected.

For calculating feed requirements it was necessary to include projections of the numbers of horses, mules, donkeys, and camels. The reasoning was as follows:

1. The downward trend in the number of horses is assumed to continue.

2. A decline is also projected in the numbers of mules, donkeys, and camels.

3. The number of the various types of livestock is expressed in terms of livestock units based upon average weights. An increase of 20 percent in total livestock units between the base period and 1985 is assumed.

Higher yields per animal would require more than proportional increases in feed availability. How much more has not been calculated. The following reflections are made about feed availability:

1. Feeding is necessary in winter except along the southern coast.

2. The pasture area is projected to remain almost stable; thus increased pasture availability has to be achieved through increased productivity on pasture lands which have been deteriorating because of overgrazing. How this can be done is not discussed in the report.

3. Hay production, which has declined considerably with the shrinking of pastures, must be increased more than 20 percent on an average. How this can be done is not discussed in the report.

4. Most of the other feedstuffs consist of straw mixed with grain. Higher yields per animal will require relatively larger quantities of grain.

5. The report concludes that livestock numbers could be expected to increase in line with the supply of cereals, provided poor grazing and haymaking conditions do not act as a brake, as happened in the 1950s.

6. The proportion of the various cereals that would be fed to livestock is assumed to be as follows: wheat, 3 percent; rye, mixed grains, and pelts, 4.5 percent; barley and oats, 100 percent; maize, 57.5 percent; and millet and canary seed, 100 percent.

According to the above estimates, the total quantity of cereals fed to livestock (including poultry) amounted to about 3.3 million tons in the base period, with barley alone accounting for about 2.2 million tons.

Feed requirements for 1975 and 1985 are projected as follows:

1. For sheep, goats, horses, mules, donkeys, and camels feed grain consumption per animal will remain unchanged. For buffalo a small increase is expected.

2. For beef and for production of cow milk feed grain consumption per kg of output will increase roughly in proportion to increases in yields.

3. The same assumption is made for poultry and eggs. However, the assumed feed grain requirements per kg of output remain below the norm of countries where industrial-type production is common. (With the large increase in production projected this may be a doubtful assumption.)

Projected feed grain requirements amount to about 4.6 million tons in 1975 and 6.4 million tons in 1985, compared with 3.3 million tons in the base period.

Consumption Projections

The consumption projections are based upon assumptions about population, and upon income growth rates and income elasticities. Table 11.1 gives the assumptions for population and income.

TABLE 11.1

Assumptions for Population and Income
Projections for Turkey
(annual rate of increase in percent)

	Population	Total Private Consumers' Expenditure	Private Consumers' Expenditure per Person
1962-65	2.4	5.2	2.8
1965-75	2.7	5.5	2.8
1975-85	3.0	5.5	2.5

Source: OECD, Agricultural Projections for 1975 and 1985; and FAO, Country Study for Turkey.

The population growth rates assumed would lead to a population in Turkey of 40.0 million in 1975 and 55.3 million in 1985. Because of rapid population growth, private consumers' expenditure per person would increase at a somewhat slower rate during 1975-85 than in the previous decade.

There are considerable uncertainties about the basic food balance data for Turkey. A balance for the base period 1961-65 was constructed by referring to various sources.

On the basis of household budgets income elasticities of demand were calculated by Roberts College (Istanbul) for rural areas and by the State Institute of Statistics (Turkey) for urban centers. The separate rural and urban elasticities were then combined into "national" elasticities. Some of these data, for various commodities, are as follows:

Commodity	National Weighted Averages
All wheats	-0.09
Rice, rice flour	0.80
Beef, veal	0.30
Mutton, lamb	1.20
Eggs	0.80
Butter, margarine	-0.12
Milk, cream	1.40
Yoghurt	0.08
White cheese	0.50

The Turkish delegation to OECD has in its comments indicated that the income elasticity for wheat should be positive in Turkey and the projected wheat consumption should be higher.

The OECD Secretariat seems to have made the projections of per capita demand in a rather pragmatic way, as can be seen from the following comments made by the Secretariat regarding per capita consumption of cereals:

> Consumption trends are likely to be influenced primarily by shifts in population from the country to the towns, which normally would mean a shift from cereals to livestock products. As yet, there is no convincing evidence that the rural population is decreasing in absolute numbers, which makes the future level of cereals consumption uncertain. At the revised overall consumption level of the base period, (even after the downward adjustment described in Annex 1) there still seems room for a decrease in per capita cereals consumption, a slight decrease in the consumption of wheat going with a large percentage drop in the consumption of rye, mixed grains and maize. Such a pattern would conform with the sign and the value of the income elasticities observed.

An increasing per capita consumption of rice has been assumed, closely related to the expansion of irrigated areas. For potatoes a positive income elasticity of 0.2 has been assumed. Sugar consumption per capita is also expected to increase, but the rate is assumed to slow down.

The income elasticities used for livestock products may seem to be high. The reasoning in the FAO report for these products runs along the following lines:

1. With rising incomes and a fast-increasing population total demand for livestock products will increase rapidly.

2. There are many obstacles to developing the livestock sector, especially shortage of feedstuffs and inadequate pastures.

3. Consequently most of the potential increase in demand for meat and dairy products would have to be met by imports.

4. For the purpose of the projections no specific assumptions on price changes, apart from existing trends, have been made.

5. A rapidly increasing consumption of poultry products has been assumed.

Overall the demand projections imply a slight decrease in per capita intake of calories and proteins. Regarding animal proteins an improvement is projected. The intake, however, will remain low by European standards. Fats, oils, pulses, vegetables, fruits, and fish are not covered by the projections.

Crop Area Projections

The reasoning in the OECD report concerning projections of crop area is as follows:

1. The total agricultural area will remain unchanged.
2. Only a small increase in cultivated area is assumed (about one million ha, or 5 percent). This implies that the decline in permanent pastures has come to an end.
3. It is assumed that the field crop area will show a slight increase, the assumption being that the fallow area will decrease only slightly.
4. An appreciable increase in the area used for horticultural crops is projected.
5. There will be no further increase for cereals totally, but a slight increase for wheat is foreseen; the area in pulses will remain unchanged; the area in industrial crops as a whole will expand; the area in oilseeds will expand considerably; and the area in fallow will remain about constant. According to the State Planning Organization (Turkey) the area in cereals has been overestimated (12.5 percent or more). In the projections a 20 percent downward adjustment has been made in the area sown for all grains, except maize and rice.

Food Balances

The projected food balances for Turkey for 1975 and 1985 are given in Table 11.2. For long-range planning the balance for 1985 is highly important. Because of the rapid growth of population in Turkey, the increasing income per capita, the rapid urbanization, and the low

TABLE 11.2

Food Balances for Major Products in Turkey, 1975 and 1985
(1,000 tons)

Product	Projections for 1975				Projections for 1985			
	Production	Balance	Total Requirements	Feed	Production	Balance	Total Requirements	Feed
Wheat	9,512	-261	9,251	285	11,620	+140	11,760	390
Rye	683	-103	580	31	707	-105	602	32
Total wheat and rye	10,195	-364	9,831	316	12,327	+35	12,362	381
Coarse grains[a]	5,338	+236	5,574	4,291	5,893	+1,361	7,254	6,017
Total grains	15,533	-128	15,405	4,607	18,220	+1,396	19,616	6,398
Rice (milled)[b]	210	+44	254	n.r.	292	+99	391	n.r.
Potatoes	2,208	-81	2,127	45	2,720	+122	2,842	54
Sugar (refined)	702	-7	713	n.r.	1,116	-5	1,111	n.r.
Beef and veal	163	+34	197	n.r.	195	+104	299	n.r.
Mutton and lamb	204	+103	307	n.r.	218	+269	487	n.r.
Goat's meat	97	0	97	n.r.	84	0	84	n.r.
Poultry meat	160	0	160	n.r.	404	0	404	n.r.
Total meat	747	+137	784	n.r.	922	+373	1,295	n.r.
Eggs	146	0	146	n.r.	256	0	256	n.r.
Milk and dairy (fat)	204	+13	217	n.r.	260	+47	n.r.	307
Products (solid nonfat)	429	-5	424	91	565	+14	579	96

[a]Mixed grains, spelts, barley, oats, maize, millet, and canary seed.

[b]At 60 percent of paddy yields.

Notes: In the "balance" columns the plus figures represent projected deficits and the minus figures represent projected surpluses.
n.r. = Heading not relevant.

Sources: OECD, Agricultural Projections for 1975 and 1985; and FAO, Country Study for Turkey.

consumption of livestock products in the base periods, large increases in demand are projected, especially for livestock products.

According to these projections, self-sufficiency in food grains can be expected in Turkey in 1985, except in rice, for which an import demand of about 0.1 million tons is projected. For livestock products, however, large deficits are projected. These deficits can be met through imports of either livestock products or the feed grains needed to produce them. If the needed imports cannot be obtained for balance-of-payment reasons, and, assuming that a ration system and/or price controls are not introduced, real prices are likely to rise, limiting the demand to the quantities available. This could create inflationary pressure. Also the future membership of Turkey in the European Common Market would limit the possibilities for import control.

The projected total deficit of meat for 1985 is about 370,000 tons. Of this deficit mutton and lamb account for approximately 270,000 tons. This deficit would occur in spite of projected imports of about 1.4 million tons of coarse grains for use as animal feed.

Poultry and egg production is expected to meet total demand, while for dairy products an import requirement of about 60,000 tons, mainly in butterfat, is projected.

The OECD report presents a pessimistic view of the prospects for expanding crop and animal yields, as can be seen from the following excerpt:

> In view of climatic and institutional conditions prevailing in Turkey, adequate increases in average yields will be difficult to obtain. Much marginal land has been ploughed up and the remaining pastures are badly over-grazed. The coarse grains area is experiencing stiff competition from the area sown to bread grains due to heavy population pressure. The prospects for providing increasing numbers of high-yielding animals with sufficient feedstuffs, therefore, do not seem very bright. Under these circumstances, it seems hazardous to project spectacular increases in yields, as it was felt that past performances should be reflected to some extent in future achievements. The major exception concerns the production of wheat where important changes

> are already under way with the introduction of Mexican wheat and better tillage practices in some regions. Milk yields per cow around the major consumption centres have also been assumed to show an appreciable increase as a result of a recently introduced livestock programme.

The OECD report is more optimistic in its comments about the industrial crops:

> The Turkish Government in its Development Plans has emphasized the importance of the agricultural sector by promoting action in many fields. The intensification of basic crop and livestock entreprises is, however, a very complex and time-consuming task under Turkish conditions. Prospects for the so-called industrial crops, oilseeds, vegetables and fruits appear more favorable as producers are better organized and accustomed to producing for an often remunerative market. Apart from sugarbeet, and potatoes no projections have however been made for these commodities.

TURKEY'S 15-YEAR PERSPECTIVE PLAN

The planning in Turkey is based upon a 15-year perspective plan covering the period from 1963 to 1977. The 15-year period has been subdivided into three five-year periods beginning with the period of the First Five-Year Plan, 1963-67.

The major long-term targets for the 15-year plan period are the maintenance of a 7 percent compound annual rate of growth; the creation of jobs for 7.5 million people in addition to the one million known disguised unemployed workers in agriculture; the improvement of income distribution; the provision of educational opportunities for all primary-school-age children; the improvement of health conditions through better health education, better nutrition, family planning, and the control of contagious diseases; the raising of domestic resources to a level where they will suffice to meet development needs, without resort to foreign assistance; and a progressive move toward a balance in external payments.

As a result of the investments and reforms to be undertaken during the 15-year period, Turkish planners expect to realize a number of structural and other changes in the economy. It is expected that the gross national product (GNP), which amounted to TL 52.7 billion ($5.9 billion) in 1962, will have increased by 175 percent to TL 145.3 billion ($16 billion) at the end of the 15-year period. At the same time, given a population growth rate of approximately 3 percent, per capita income should increase by 78 percent to TL 3,200 ($350). During the 15-year period the agricultural sector's share of GNP is expected to decrease to less than 25 percent, while the share of the industrial sector should increase to about 30 percent. In this period the proportion of active population engaged in the agricultural sector is expected to fall from 77 percent to 58 percent, while the proportion of active population in the industrial sector is expected to increase from 10 percent to 16 percent.

The OECD agricultural projections cited above seem to be inconsistent with some of the long-term targets of the Turkish perspective plan, including the maintaining of a 7 percent compound annual growth rate of GNP; the creation of full employment; and the move toward a balance in external payments.

The Turkish delegation to OECD also commented that the assumptions used for the OECD projections are not consistent with the high GNP growth rate targeted in the Turkish plan. The 5.5 percent increase in private consumer expenditure projected by OECD would probably correspond to about 6 percent growth in GNP. A higher growth rate would result in higher demand, especially for livestock products, than that projected by the OECD Secretariat, and in larger deficits assuming the same growth in agricultural production.

The growth rate in agricultural production might, however, be higher. The assumptions for the agricultural production projections of OECD represent, more or less, a "status quo philosophy" based upon prevailing trends and tendencies. This can be illustrated by the following examples: that only moderate increases in yields of field crops are projected because it is assumed that "improved varieties and fertilizers are of little avail"; by the assumption that the fallow area is to remain about constant; and by the projected increase of livestock production representing only what can take place without improved technology.

It seems obvious that with the growth in food production assumed in the OECD projections large deficits may occur. These deficits would place a heavy burden on the Turkish economy, especially regarding the balance of payments and the risk of inflationary pressure. Turkey's gradual integration into the EEC could also become difficult. Hopefully it will be possible to increase food production more rapidly through new technology and improved strategy for agricultural development. However, the situation raises the following difficult questions: (1) Is it possible, through the introduction of proper packages of new technology, to increase agricultural production much more rapidly than projected in the irrigated areas? (2) Can the same increase take place in the rainfed areas, through the development and introduction of new grain varieties and improved cultural practices? (3) Is a significant reduction in fallow land possible? (4) Can wheat yields be increased much more rapidly than assumed, making it possible to divert land from wheat to other crops, first in the coastal areas and later in the plateau? (5) Can a new system of mixed farming based on wheat-fodder-sheep production (as in Australia) be introduced, together with sheep breeds with higher fertility, thereby facilitating a much more rapid increase in the production of mutton and lamb? (6) Can intensive feed production be expanded in higher rainfall and irrigated areas, for example by using sorghum or maize, facilitating a much more rapid increase in the production of poultry, beef, and veal? (7) Can the introduction of urea, and chemicals for the treatment of straw, increase the feeding value of roughages and help to provide feeding rations with sufficient protein content? (8) Can the impact of irrigation water for which development is planned or under implementation on an additional 2-3 million hectares, and improved water management, be much more pronounced than was assumed in the OECD report? (9) How much reclamation of the potential 5 million hectares of potential arable land reported by Sommetz (see Chapter 2) can be carried out?

Only well-planned and vigorously implemented research can answer these and other questions. The tentative conclusions must be that the potential for agricultural development in Turkey through 1985 is much greater than indicated by the OECD report, and, given a proper strategy, this growth can be much more rapid: This situation represents a great challenge.

CHAPTER

12

INTRODUCTION OF THE MEXICAN WHEAT VARIETIES

INITIAL TESTING OF THE MEXICAN WHEAT VARIETIES

Wheat has always been the most important crop in Turkey in terms of both acreage and production, and it is likely to keep its relative importance. Wheat is at present sown annually on approximately 9.0 million hectares, of which some 0.4 million hectares of area receive one or more irrigations.

There are two main sets of climatic conditions in Turkey under which wheat is grown. The first set of conditions includes limited rainfall (under 400 millimeters per year), cold winters, and little, if any, developed irrigation water. Under these conditions, which characterize much of the Anatolian Plateau, winter wheat is grown. The second set of conditions includes generous rainfall, a mild climate, the availability, in many cases, of irrigation water. These conditions are found along the coast. It was recognized at a relatively early date, by wheat scientists, that the coastal areas were most likely to be suitable for cultivating the Mexican wheat varieties.

Wheat yields in the coastal regions of Turkey were practically stagnant during the 1960s until the introduction of the Mexican varieties. Previous production increases were mainly a result of increased acreage. The stagnation in the yield level resulted from a lack of new high-yielding varieties, and from the continued use of traditional cultivation methods, inadequate seed bed preparation, mainly broad casting of seed and fertilizer, and no significant increase in the use of fertilizer. The situation was ripe for introduction of new high-yielding technology.

The testing of the Mexican varieties started relatively early in Turkey, two to three years earlier than it started in the Indian subcontinent. In 1959 Sadittin Demiröz, Director of the Turkish Agricultural research station at Adapazari, contacted Norman E. Borlaug, Director of the International Wheat Program, and his coworkers in Mexico. Demiröz obtained 17 samples from the Mexican breeding stocks. These were all planted in Turkey in the 1960 season. More material, including the released varieties, was obtained in subsequent years. In 1960 and 1961 several Turkish agricultural scientists visited Mexico, and had an opportunity to observe the work that was going on with a new type of high-yielding wheat. The scientists brought back several new varieties, which were placed in the hands of Turkish wheat breeders and planted at research stations at Adapazari and Eskisehir. It was found that the climate in Eskesehir is not suitable for the Mexican varieties, but in Adapazari the Mexican varieties were grown for several years, and they out-yielded other wheats, collected from many countries, when grown in conjunction with high levels of moisture and fertilizers. However, no program for the introduction of the Mexican varieties was considered in Turkey either by the wheat breeders or by the government until 1967, despite pressure exerted by farmers interested in the Mexican wheats, and despite the successful introduction of these varieties in the Indian subcontinent. By 1967 these varieties had been tested for six years, and four Turks had over the years received one year's training in Mexico in wheat and barley breeding.

In a report of 1966 by Abdul Hafiz, it is noted that a Mexican wheat variety, 8156, which was successfully introduced in India and Pakistan, gave a yield of 6.5 tons per hectare in a microplot trial for the 1965-66 season at Adapazari.* More than 800 other varieties brought in from Mexico were screened during that season at the Adapazari station. The impression one gets from this report is that instead of directly introducing the Mexican varieties, the Turkish wheat scientists regarded it better to cross the best of the Turkish varieties with the Mexican varieties. Even if successful, such a program would, however, take a long time. It turned out that the farmers were not willing to wait.

---

*Abdul Hafiz, "Report on Wheat and Barley Improvement in Turkey," FAO, Near East Wheat and Barley Improvement and Production Project (Cairo, August 1966).

Testing and Spearheading by Farmers

In the summer of 1965 a progressive farmer in the coastal Cukurova area, Mehmet Can Eliysil, learned about the Mexican wheats and their successful introduction in Mexico, Pakistan, and India. He obtained, through an AID technician, 15 kilos each of two Mexican varieties, Sonora 64 and Lerma Rojo, from India. He planted the seed in the fall of 1965 on his farm, near Tarsus, and had the foresight to put nitrogen fertilizer on the crop. His experiment turned out to be of great importance in later events. He harvested more than 4.0 tons per hectare, or about double the yields of the best local wheats in the area. The excellent performance of these Mexican varieties caught the interest of other farmers in the Tarsus-Adana area. Some 106 progressive farmers of the Cukurova Chamber of Agriculture pooled their money and requested permission from the central government to import 60 tons of Sonora 64 wheat seed from Mexico for planting in the fall of 1966. After considerable parliamentary discussion the farmers received permission and the necessary foreign exchange to purchase the wheat. The cost of the seed, delivered at Iskenderun, was equal to $259 per ton, which, together with customs and other costs, represented a sizable investment for the farmers. Because the seed arrived late, planting was not completed until December. The farmers were encouraged by U.S. AID officials to use adequate quantities of nitrogen fertilizer with the seed: most of the farmers did this and the results were generally highly satisfactory. Large numbers of other farmers came from considerable distances to view the new wheat and became interested in it. This resulted in strong demands by a group of farmers from Adana Province for imports of large quantities of the Mexican seed for planting in the fall of 1967. This group spearheaded the initial program for the introduction of the Mexican wheats in the coastal areas of Turkey. In the meantime the minister of agriculture had taken action.

Worried about the lagging agricultural production in Turkey, the Minister of Agriculture, Bahri Dagdas, requested AID to provide a team of agricultural specialists to visit Turkey from the middle of November to the middle of December of 1966 to study Turkey's agricultural situation and make recommendations for its improvement. For the future introduction of the Mexican varieties it was a happy coincidence that the six-man team chosen was headed by Charles M. Elkinton, who was then the food and agriculture officer for AID in Pakistan, and who had supported the introduction of the Mexican varieties in Pakistan. Naturally the team looked for opportunities for the rapid introduction of Mexican wheats in Turkey; and they found them in the coastal areas. They concluded that it should be feasible to adopt a number of the high-yielding Mexican spring wheat varieties for these well-watered and climatically mild areas of Turkey. They recommended an approach based on a large number of dramatic demonstrations of a package consisting of high-yielding varieties and improved cultural practices, which had proved to be effective elsewhere, in conjunction with an appropriate support price to encourage wheat producers to adopt new varieties and improved production practices.

While the team recommended that Turkey introduce, on an extensive scale, high-yielding Mexican wheat varieties, it did not advise the government to undertake a large-scale seed import program for the 1967-68 season. Instead the report included a suggestion that a team of cereal specialists study the performance of the Mexican wheat being grown by the 106 farmers in Cukurova and make recommendations on the possible introduction of Mexican wheat varieties on a larger scale. Since the Elkinton report was submitted in December 1966, it would have been too late to arrange for large-scale seed imports if the government had waited for the report from the team of wheat specialists in the spring of 1967. The Turkish minister of agriculture understood this, and, although accepting the recommendation regarding the team of cereal specialists, he decided not to wait, but to launch the import program.

Serious Opposition to the Program from the Research Scientists

Turkey was lucky to have a man like Dagdas as its minister of agriculture, during the major introduction of the new wheat varieties. He is considered a very strong-willed individual and he placed top priority on this project, which meant that much of the red tape associated with such a project was removed, and that serious opposition to it could be overruled. He decided, on the basis of experience already gained at the research stations and by the Cukurova farmers, and after being pressed by these farmers, to initiate a crash program of importing large quantities of Mexican wheat seed. He suggested imports of 50,000 to 60,000 tons and asked AID to provide long-term credit for the purchase of the seed.

In January and February many meetings were held to discuss the feasibility of the program, and how to carry it out: Turkish research personnel strongly opposed the program on the basis that not enough was known about the adaptation of the Mexican varieties. They wanted to test the varieties for several more years before eventually approving large-scale imports of the seed. The regulations in Turkey require that before a new wheat variety can be approved for introduction it must be grown for at least five years and meet the standards set by Turkish scientists. This rule had to be waived for the importing of the Mexican varieties.

Serious doubts about the program were also expressed by some officials of the AID Mission in Ankara. They questioned the proposed size of the program, the ability of the extension service to support it, the dangers of rust and pests, the lack of machinery to harvest the Mexican wheats, and the baking quality of the wheats.

The determination of the minister (Dagdas) and a firm stand in favor of the program by the chief of the agricultural branch of the Near East and South Asia Bureau of AID in Washington, O. L. Mimms, were the main factors leading to the decision that the program was considered feasible, and was to be carried out. However, further discussions among the parties involved resulted in a reduction of the request for Mexican seed from the original 60,000 tons to 20,000 tons.

At the request of the Turkish Ministry of Agriculture, AID, in the spring of 1967, arranged for three U.S. wheat consultants to visit the Cukurova region of Turkey, where they could inspect the Mexican Sonora 64 seed being grown there. The consulting team was composed of Orvel Vogel, of Washington State University, and Tom Jackson and Warren Kronstad, both of Oregon State University. The three men made two three-week visits to the Cukorova region that were geared to coincide with the heading and harvesting stages of the Mexican wheats. A preliminary report submitted by Jackson and Kronstad after the first visit, gave a rather lukewarm report regarding the program. This alarmed AID in Washington, which again raised the question of the advisability of undertaking the importing program. It appears that Kronstad and Jackson, to a certain degree, shared the view of the Turkish wheat scientists that the Mexican varieties had not been adequately tested, and that the program involved a considerable disease risk.

AID in Washington also questioned whether Turkey could maintain sufficiently close control over at least one-fourth of the total planting, which AID had set as a condition for support and financing. Again, the program faced a crisis. Assurance was given that the State Farm organization was fully capable of maintaining the requested control, and while the discussions continued the letter of credit for purchase of the seed was received in Mexico on April 19, and the work proceeded.

By March 8, the government of Turkey sent a group of three wheat research technicians to Mexico to join a Turkish technician, who was in Mexico under training, to inspect the fields from which Mexican seed was to be obtained, and observe the harvesting, cleaning and bagging of the seed. Together the four men had the full responsibility for inspection and approval of the seed. They received advisory aid on the seed purchase from Norman E. Borlaug and Charles F. Krull of the Rockefeller Foundation, and Ralph Gleason and Robert Horton of AID. Working together the Ministry of Agriculture and AID decided which varieties were to be imported and made other important decisions.

Originally the plan had been to purchase 5,000 tons of certified seed, to be planted in Turkey under the supervision of state farms, and 15,000 tons of commercial-grade seed for distribution to Turkish farmers. However, there were only 1,770 tons of certified seed available; and the amount of commercial-grade seed was increased to 19,930 tons. The cost of the commercial seed was approximately $120 per ton bagged, FOB, and the cost of the certified seed was approximately $150 per ton bagged, FOB. The tonnages that were obtained in the two varieties are shown in Table 12.1.

Strategy for Introducing Mexican Wheats

The strategy involved in the program for introducing the Mexican wheats in Turkey represented some important changes in the procedures of the Turkish government. An Inter-Ministerial Wheat Council was established to determine the policies for the program and to assure a coordination of efforts. Four wheat committees were organized, one in the Ministry of Agriculture in Ankara, and one in each of the three major coastal regions suitable for adoption of the Mexican wheats, namely the Mediterranean, the Aegean and the Marmara region. The Ministry of Agriculture committee, chaired by Ekrem Günay, assistant general director of agriculture, was charged with the implementation of the entire program. The regional committees, headed by the directors of the agricultural research institutes in the three regions, were charged with conducting the provincial and county programs in their respective regions. The minister of agriculture insisted on complete cooperation in the program.

TABLE 12.1

Quantities of Mexican Varieties and Grades of Wheat Seed Imported by Turkey in 1967

Seed	Certified Grade	Commercial Grade	Total
Penjamo 62	600	6,347	6,947
Lerma Rojo 64	800	5,390	6,190
Super-X	0	6,856	6,856
Mayo 64	270	915	1,185
Nadadores 63	60	435	495
Sonora 63	20	0	20
Pitic 62	0	320	320
Inia 66	5	0	5
Noroeste 66	5	0	5
Tobari 66	5	0	5
Oviachic 66	0.3	0	0.3
Jaral 66	4.7	0	4.7
Ciete Cerros 66	1.0	0	1.0

Source: Food and Agriculture Division (AID: Turkey).

Few Turkish farmers or extension workers were acquainted with the new Mexican wheats, or with the cultural practices required to grow them; an extensive educational program was developed to assure success. The Turkish government assigned 250 extension agents to the coastal regions to work full time on the Mexican wheat program. AID, at the request of the Turkish minister of agriculture, agreed to provide a team of 12 extension agents from the states of Washington and Oregon, on the basis of a contract with Oregon State University, with nine of the team members working on the extension program for three months and three members working for one year. The AID mission in Turkey helped to finance the purchasing of 50 trucks and 145 pickups, as well as the drivers' salaries, per diem expenses for the drivers and the extension agents, and the development and publication of educational, training, and publicity materials. The following materials were developed to instruct both the extension agents and farmers:

	Quantity
Leaflet entitled "Why Mexican Wheat"	25,000
Booklet containing instructions to agricultural extension agents	400
Bulletin entitled "How to Grow Mexican Wheat"	45,000
Flip chart showing the major steps	300
Film slides (40 slides in each set)	40
Record cards for farmers and county agents	60,600
Poster emphasizing the need for drainage	5,000
Poster emphasizing spring fertilizer applications	15,000
Folder urging correct fertilizer use	50,000
Poster urging proper preparation for the second year	5,000
Reprint of "How to Grow Mexican Wheat"	110,000

The following basic principles for growing Mexican wheat were emphasized in the printed material and in the extension work:

1. Prepare an appropriate seed bed.
2. Plant between November 1 and December 20.
3. Use eight kilograms of seed per decare.
4. Apply six kilograms of phosphate per decare at planting time.
5. Apply four kilograms of actual nitrogen at planting time, four kilograms in January, and four kilograms in February.
6. Use seed and fertilizer drills.
7. Make drainage canals where they are necessary and keep the field drained.
8. Eradicate weeds.
9. Follow all of these steps to obtain a yield of 350-600 kilograms per decare compared to 100-200 kilograms per decare for Turkish wheat.

In order to train the extension agents a three-day seminar was planned for each of the three coastal regions in which the Mexican seeds were being planted--Mediterranean, Aegean, and Marmara.

The American extension team arrived in Ankara in August 1967. The team was strengthened through the addition of four AID technicians who had been evacuated from Middle East countries, because of the June 1967 six-day war. One of these four technicians, L. M. Humphrey, was appointed chief of the Cereals Production Branch of AID in Turkey, and put in charge of the AID aspects of the project. The others were assigned liaison duty in the three coastal regions, increasing the number of Americans working directly in the field to 15.

In addition 15 Turkish officials, representing the extension service, state farms, and the Ministry of Agriculture, were sent on a three-week mission to study wheat production in the United States (the Pacific Northwest) and Mexico. Vogel, Jackson, and Kronstad guided the Turkish officials during their tour.

Those aspects of the program involving the provision of credit and fertilizer inputs were planned well in advance. The fertilizer was furnished through imports, and was distributed by the Agricultural Supply Organization and by the credit cooperatives. The required credit was made available to the farmers through the Agricultural Bank.

In carrying out the program the American extension team initially encountered two problems: inadequate transportation and the need for interpreters (which the Turkish government had been expected to fulfill). However, both problems were quickly solved when AID rented cars for the team and hired interpreters.

The Turkish counterparts of the American personnel were the directors of agriculture in each province. These Turkish officials worked closely with all extension personnel on the program, although a few of the directors showed indifference in their overall attitude toward the program, which at times led to a strain in relations at the higher levels. At the lower levels, where most of the major work was done, the coordination was consistently effective.

The program represented a significant change for the Turkish extension workers, who had not been accustomed to working closely with the farmers--generally they had spent most of their time on administrative work. They were poorly trained in the required subject matter, and had little knowledge of how to effectively train farmers. For the wheat program, however, they were relieved of their extraneous duties, plus the fact that they received detailed instructions in the production of the Mexican wheats at the regional seminars, and were given the variety of printed materials described above. Also they were well supported by their American counterparts. Together the Turkish and American teams developed what has been called "the teahouse approach": Turkish farmers tend to meet at the village teahouse during periods when there is no field work. This provided a good opportunity for the extension agents to discuss the program with farmers in an informal atmosphere. Frequently the agents and farmers gathered at the village teahouses several times each day. The farmers' response to these meetings was so enthusiastic that more than 60,000 of them took part in the program. Moreover, every one of these farmers was personally instructed by an extension agent at least once, and most of them received instruction several times.

The extension agent's job was to promote not only the new Mexican wheat varieties, but better seed bed preparation, a substantial reduction in the seeding period, the use of grain drills instead of broadcasting methods, higher rates of fertilizer application at the proper times, and weed control. In this respect the Turkish program was perhaps more complete than any other national program for the introduction of Mexican wheats. The extension effort was also more concentrated than for any other program. The Turkish government did not pay any subsidy for the program. The price of wheat was supported at the same level that was in effect before the introduction of the Mexican wheats (75 kurus per kilogram).

The farmers were required to pay the full costs of the seed, fertilizer, and drills for the program. However, credit was made more readily available to farmers than in the past, with the following system instituted: Wheat loans were granted on a note with two signatures, instead of requiring a mortgage on the land (as in the past); loans were approved even though a previous crop loan, on cotton for example, had not been repaid by a farmer; credit at 7 percent interest, for the purchase of drills or other equipment, was made available on the basis of two, three, or four signatures; in case of a disaster, causing the loss of a crop, loans were extended for a year and a second wheat loan was granted; branches of the Agricultural Bank sent their personnel to distant villages to help farmers expedite applications for their loans; all loans were "in kind," and to obtain credit for seed the farmers were required to use the recommended amount of fertilizer.

The key practices involved in growing the Mexican wheats in Turkey were so different from the traditional Turkish wheat practices that some farmers found them difficult to accept, or they often did not have the means required to carry out each step. Some of the problems that developed as a result of these differences were the following: improper seed bed preparation, resulting in lumpy, uneven seed beds; untimely planting; uneven seed broadcasting, resulting in a noticeably uneven stand of wheat; uneven planting depth (where broadcasting was used seeds were not evenly covered at the required depths); uneven fertilizer application, where broadcasting was used, resulting in uneven plant development; and irrigation water was often not applied at the most critical time, during the ground root development.

In many areas there was also a shortage of drills, and a lack of maximum utilization of the drills available. The very nature of the Mexican wheat necessitates planting with a drill at a depth of five centimeters to insure a maximum yield. Broadcasting results not only in uneven spreading of the seeds, but also uneven covering of the seeds.

To provide transportation for the Turkish extension staff, about 50 vehicles were transferred on a temporary basis from other governmental agencies; about 145 vehicles were secured from U.S. military surplus sources; and the importing of vehicles was speeded up.

Adverse publicity on the Mexican wheat program began to appear in the press even before the new seeds were planted.

> "Mexican wheat is worm-infested. . . ."
>
> "The flour derived from Mexican wheat, which is sprayed with poisonous chemicals, spreads death. . . ."
>
> "America is exploiting this country. . . . By feeding Turks on cereal regime, the U.S. is sponsoring the production of so many stupefied Turks. . . ."
>
> "The U.S. is recommending Mexican wheat exclusively with the aim of sabotaging cotton production in Turkey. . . ."

Most of this publicity was originated by leftists who opposed the use of outside assistance to improve Turkey's agricultural situation. Mostly this propaganda was counteracted by vigorous action from the Ministry of Agriculture.

Approximately 17,000 tons of the new Mexican seed were planted in Turkey in the 1967-68 season. The area covered was about 165,000 hectares, with about 69,000 hectares being located in the Mediterranean region, 35,000 hectares in the Aegean region, and 61,000 hectares in the Marmara region. The average yield for the entire area of Mexican wheat was 3.5-4.0 tons per hectare, in spite of relatively unfavorable weather conditions. Using a conservative estimate for yields of Mexican wheats and a relatively high figure for yields of traditional varieties, the figures would indicate nearly 0.6 million tons of total production of Mexican wheats and a minimum of nearly 0.35 million tons of additional wheat output. At the world price prevailing at the time of harvest the crop was worth more than $23 million. Total costs for the program during the first year were estimated at approximately $15 million, leaving a net profit of $8 million. In addition there were many other favorable results of the program, including a rapid expansion in the manufacturing of grain-fertilizer drills and other equipment. A study of 154 farms by the Ministry of Agriculture showed that drilled seed yielded approximately 0.8 tons per hectare more than seed broadcast by hand.

Data from the Agricultural Bank indicate that 80 to 85 percent of the farmers growing Mexican wheat obtained credit. Very little grain was offered for sale to TMO from the 1967-68 Mexican wheat crop.

Approximately 70 percent of the farmers participating in the program during the first year might be classified as small farmers, and 30 percent might be regarded as relatively large farmers. At least 70 percent of the Mexican wheat was harvested with combines. There was a transfer of combines among areas that had different harvest seasons.

EXPANSION OF MEXICAN WHEATS

A wide expansion during the second year (1968-69) of the program increased the Mexican wheat area from approximately 165,000 hectares to 770,000 hectares. The number of farmers involved increased from 60,000 to more than 200,000, and the new seeds were tried in nearly every province of Turkey, as illustrated in Figure 6. Three provinces in southeast Anatolia: Urfa, Diyarbakir, and Mardin, were brought into the program with substantial plantings of the Mexican seeds in the fall season, and in the other 25 provinces in the Anatolian region small plantings were made on irrigated land in the spring season. The entire 1968-69 campaign was carried out by the Turkish government. There was no increase in extension personnel, facilities, or financing, and consequently the highly concentrated job that was done in the coastal areas during the first year of the program could not be repeated. Even in the new areas where the Mexican varieties were introduced, the efforts to spread the new growing practices slowed down; however, the improved administrative practices helped to offset this. The potential for tripling wheat yields in the coastal areas represented a great stimulus for increasing the average acreage devoted to the new varieties.

The economic results of the Mexican varieties were also highly favorable. A study by the Agricultural Planning and Economic Research Department of the Turkish Ministry of Agriculture, shows that net income from land and management per hectare increased in the first year about 150 percent with the introduction of the new technology. The results of this study are shown in Table 12.2.

A number of problems arose in the second year of the program, including a buildup of wheat diseases (stripe rust and septoria), which had been feared by the Turkish wheat scientists and the American consultants from the very beginning. The high average yields per hectare for the 1967-68 season were due in part to a relatively disease-free year. As indicated in Table 12.3, the average yields for the 1968-69 season were considerably lower than for the previous season.

Contributing to this decline was the large increase in the number of farmers who were not given a concentrated educational program; a shortage of nitrogen fertilizer; and the frost that occurred in the late spring. The average Mexican wheat yield in the 1968-69 season was about 2.4 tons per hectare, compared with a wheat yield of 1.2 tons per hectare for Turkey as a whole.

The results of the plantings of Mexican wheats in the noncoastal areas did not meet initial expectations; and thus the Mexican varieties have been generally limited to the coastal areas. The expectations, following the successful 1967-68 season, for very rapid and widespread acceptance of Mexican wheats (to the extent of covering 1.3 million hectares) did not materialize.

FIGURE 6

Planting of Mexican Wheats in Turkey, 1968-69

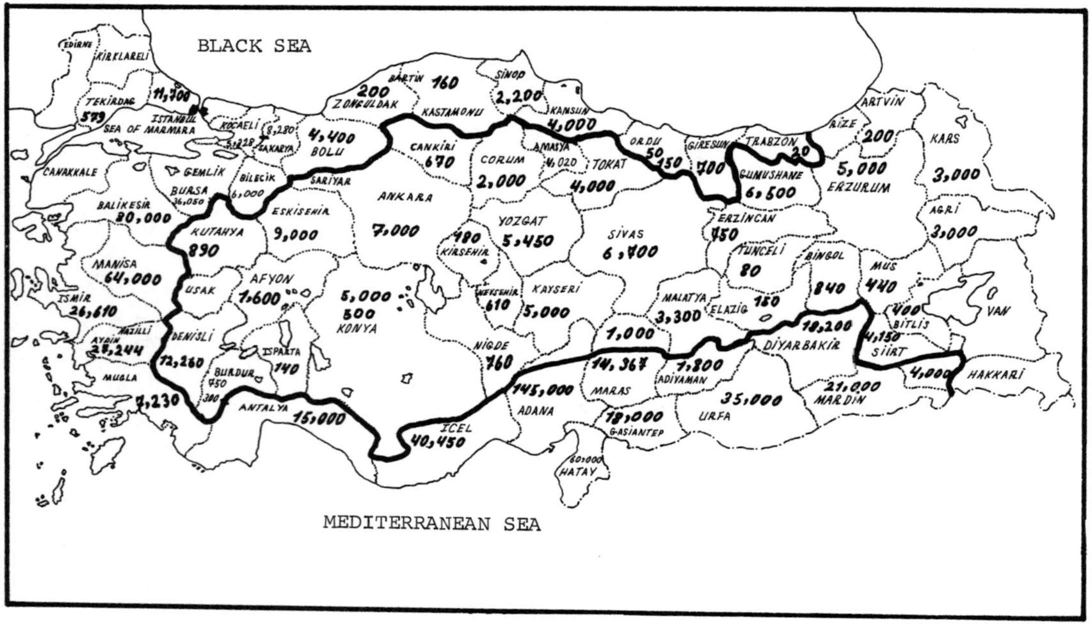

Provincial map of Turkey showing the number of hectares of Mexican wheat planted in each province in 1968/69. Those inside the heavy line were planted in the spring, those outside, in the fall. Sixty one provinces had Mexican wheat in 1968/69 and six had none.

Source: AID--Turkey.

TABLE 12.2

Income and Cost Comparisons of Mexican Wheat and a Native Variety Grown in Denisli Province of Turkey, 1967-68

	Mexican Wheat	Native Wheat
Number of farms	55	44
Hectares per farm	3.2	2.5
Costs per hectare (dollars)		
Family labor	3.54	2.60
Hired labor	7.56	5.27
Seed	18.33	16.90
Water[a]	3.62	0.87
Fertilizer	37.58	12.56
Power and implements	46.79	32.71
Interest (7.0 percent)	8.22	4.96
Total[b]	125.64	75.87
Production and income		
Grain (kilograms)	3,643	1,572
Price per kilogram (in 10 kurus)	8.88	9.89
Total value (dollars)	323.44	155.43
Net income (dollars)[c]	197.80	79.56

[a]Most of the producers did not use irrigation for wheat.
[b]Not including land rental.
[c]From land and management.

Note: Data for Mexican wheat are simple averages for three varieties: Super-X, Lerma Rojo 64, and Penjamo 62. The native variety (073-44) is one of the better native varieties for this area. Dollar values are converted from Turkish lira at nine TL per dollar.

Source: Agricultural Planning and Economic Research Department, Ministry of Agriculture, Turkey.

TABLE 12.3

The Contribution of Mexican Varieties to Turkey's Total Wheat Production

Year	Total Wheat Area (thousand hectares)	Mexican Wheat Area (thousand hectares)	Mexican Wheat as Percentage of Total Wheat Area	Total Turkish Wheat Production (million tons)	Production of Mexican Wheat (million tons)	Average Yields of Mexican Wheats (tons per hectare)	Mexican Wheat as Percentage of Total Production
1966-67	8,000	n.s.	n.s.	10.0	n.s.	4.0	n.s.
1967-68	8,250	170	2.2	9.5	0.54	3.2	5.7
1968-69	8,660	650	8.9	10.5	1.84	2.4	18.3
1969-70	8,600	650	8.2	10.1	1.96	2.8	23.1
1970-71	9,000	509	5.7	12.6	1.60	3.1	12.7
1971-72	9,000	623	6.9	11.0	1.80	2.9	16.4

Note: n.s. = Not significant amount.

Source: Unofficial estimates by the Wheat Research and Training Project.

For the 1969-70 season there was no increase in the area devoted to Mexican wheats, partly because of the stripe rust attack during the previous season and partly because of heavy rains in some of the coastal areas, which prevented extensive field activity. Most of the area for Mexican wheats in the 1969-70 season was planted with Penjamo 62 and Lerma Rojo 64. A fungus disease (Septoria Blight), also caused yields to fall below expectations, particularly in the Mediterranean region. The conservatively estimated loss for the season was 10 percent of the expected yields. In spite of this loss average yields were higher than for the previous season.

Partly as a result of the fungus disease, the area of Mexican wheats declined for the 1970-71 season from 650,000 hectares to slightly more than 500,000 hectares. In addition, of the several imported Mexican wheats, essentially only one variety--Penjamo 62--was being grown in the coastal areas. This variety has moderate resistance to stripe rust, septoria, and leaf blotch. Penjamo 62 is now being grown on about 90 percent of the Mexican wheat area in Turkey. Another reason for the lower acreage of Mexican wheats in the 1970-71 season was a higher cotton-wheat price ratio. The ratio between cotton and wheat prices increased from 5.8 in 1969 to 7.1 in 1970, as indicated in Table 12.4.

For the 1971-72 season the area devoted to Mexican wheats again increased to above 600,000 hectares, in spite of the fact that the relative cotton price became still more favorable for cotton. The future price relationship between cotton and wheat will be an important factor influencing the cropping pattern in some of the irrigated coastal areas. Data from a 1969 study by the Ministry of Agriculture indicate that, at the price relationship prevailing in 1967 and 1968, Mexican wheat was more profitable than Turkish cotton where the yield of cotton was less than 200 kilograms of seed cotton per decare. Some people were afraid that this would result in a decline in cultivated area devoted to cotton, which would have been undesirable. However, the cotton area did not decline, and subsequently showed a considerable increase.

In some areas, such as the Cukurova Plain, the introduction of the Mexican wheats had a spectacular impact. This area, which had not produced very much wheat previously, suddenly became a key center for wheat production. Wheat purchases by TMO in Cukurova in 1967 were only around 47,000 tons. Purchases rose to 55,000 tons in 1968, to 141,000 tons in 1969, 281,000 tons in 1970, and 444,000 tons in 1971.

TABLE 12.4

Price Ratios in Turkey between Competing Products and Wheat

Year	Dry Beans-Wheat Ratio	Cotton Lint-Wheat Ratio	Alfalfa-Wheat Ratio	Barley-Wheat Ratio	Sunflower-Wheat Ratio
1961	2.2	7.1	0.44	0.74	1.5
1962	2.1	6.4	0.43	0.72	1.4
1963	2.8	6.5	0.40	0.72	1.6
1964	2.9	6.4	0.40	0.72	1.7
1965	2.5	6.0	0.41	0.76	2.0
1966	2.3	5.7	0.41	0.79	1.9
1967	2.3	6.0	0.43	0.79	2.2
1968	2.6	6.0	0.44	0.81	1.9
1969	2.8	5.8	0.50	0.83	1.8
1970	3.0	7.1	0.53	0.69	1.9
1971	3.4	7.3	0.51	0.83	2.0

Source: State Institute of Statistics. Ratios calculated by the author.

As shown in Table 12.5, Adana Province as a whole (of which the Cukurova Plain is a part) has led the way in the introduction of Mexican wheats. In the 1969-70 season Adana alone had 2.5 million decares devoted to Mexican wheat, or about 46 percent of the total Mexican wheat area in Turkey.

In wheat yields for areas in which the new wheat technology has been introduced, there is also a striking difference in yield between Adana Province and the other provinces, as shown in Figure 7. For Adana the average yields were 3.2 tons per hectare in 1971, compared with a level of 1.8-2.0 tons per hectare for the other provinces. Previous to the introduction of the new technology average wheat yields for Adana were approximately at the same level as for the other areas: from 1.4 tons per hectare. As pointed out previously, it was in Adana that farmers first got hold of Mexican wheat seeds and later urged the government to negotiate a substantial seed import. The pioneering spirit of the farmers in this area still seems to be very much alive: the province of Adana

TABLE 12.5

Area of Mexican Wheats in Turkey, 1968-71
(decares)

	1971 Harvest for Various Mexican Varieties								Other Total Annual Mexican Wheat Harvests		
Province	Penjamo	Lerma Rojo	Pitic	Inia	Tobari	Super-X	Nadadores	Total Mexical	1970	1969	1968
Adana	1,768,867	105,100	4,328	5,180	n.s.	n.s.	n.s.	1,883,475	2,508,700	1,778,165	396,020
Antalya	230,563	11,500	n.s.	n.s.	n.s.	5,500	n.s.	247,563	222,090	150,000	29,920
Gaziantep	48,608	1,225	n.s.	1,500	900	15,000	1,000	68,233	180,000	180,000	31,540
Hatay	398,000	n.s.	n.s.	n.s.	2,000	n.s.	n.s.	400,000	420,000	500,000	115,240
Icel	238,085	n.s.	n.s.	n.s.	n.s.	n.s.	n.s.	238,085	520,000	400,000	70,760
Maras	81,530	27,450	n.s.	n.s.	n.s.	9,500	n.s.	118,480	55,000	164,000	43,410
Aydin	112,900	28,820	400	n.s.	n.s.	80	n.s.	142,200	109,320	258,000	81,180
Denizli	70,400	40,600	n.s.	n.s.	n.s.	n.s.	n.s.	111,000	120,000	200,000	36,980
Izmir	110,539	59,691	n.s.	n.s.	n.s.	n.s.	n.s.	170,230	152,711	237,000	100,840
Manisa	353,011	101,356	142	n.s.	n.s.	50	n.s.	454,559	440,000	600,000	108,900
Mulga	53,140	14,910	n.s.	n.s.	n.s.	n.s.	n.s.	68,550	69,870	72,000	22,630
Balikesir	123,219	8,700	13,990	n.s.	n.s.	10,151	8,000	164,060	138,410	265,000	111,300
Bilecik	17,640	15,000	n.s.	n.s.	n.s.	n.s.	n.s.	32,640	30,000	51,000	2,000
Bolu	47,265	3,000	n.s.	n.s.	n.s.	n.s.	500	50,765	35,000	44,000	7,760
Bursa	84,580	3,248	n.s.	n.s.	n.s.	45	n.s.	87,873	119,660	360,000	83,640
Canakkale	117,115	2,833	13,667	n.s.	n.s.	2,000	n.s.	135,615	62,040	115,311	58,850
Istanbul	15,725	600	1,500	n.s.	n.s.	1,100	300	19,225	35,375	107,626	43,300
Kocaeli	16,665	635	n.s.	n.s.	n.s.	n.s.	n.s.	17,300	25,990	45,000	12,490
Sakarya	55,980	11,330	n.s.	n.s.	n.s.	3,900	n.s.	71,210	72,550	275,000	71,760
Samsun	310,000	n.s.	n.s.	n.s.	n.s.	n.s.	n.s.	310,000	120,000	40,000	
Total	4,254,332	435,998	34,027	6,680	2,900	47,326	9,800	4,791,063	5,436,716	5,842,102	1,428,520

Notes: One decare is equal to 0.2471 acres; figures are given for the 20 provinces that have stayed in Mexican wheat production. In some former years provinces such as Tekirdag, Diyarbekir; Mardin, Urfa, and several Anatolian plateau provinces were included in total area calculations, but these areas were not included in the Mexican wheat program in 1970 and 1971. n.s. = Not significant amounts.

Source: Food and Agriculture Division, AID, Turkey.

FIGURE 7

Wheat Yields for the Main Regions of Turkey in Which New Wheat Technology Has Been Introduced

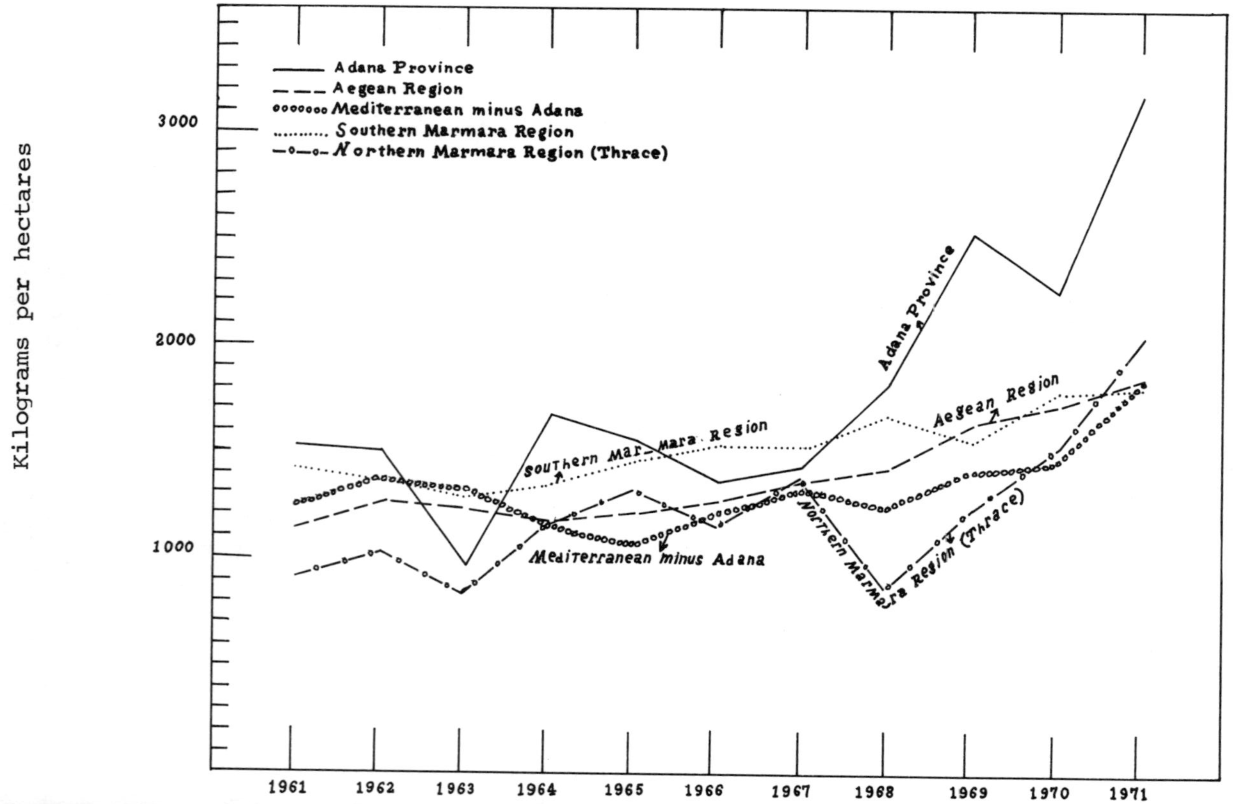

alone currently shows higher wheat production than each of two of the nine Turkish agricultural regions, and is approaching the same production level as that of three other regions. The relatively more rapid increase in the cultivated area, yields, and production of Mexican wheats in Adana can perhaps partly be explained by the fact that agriculture in this province is highly commercialized and mechanized. Also the farming units are relatively large, and Adana has a Turkish- and World Bank-supported program for land leveling and the introduction of modern irrigation methods.

Adana is the only province where the Mexican varieties of wheat are dominant, with these seeds covering nearly 84 percent of the total wheat area. The provinces of Hatay and Manisa have between 40 and 50 percent of their wheat area devoted to Mexican varieties; Samsun, Icel, and Antalya have between 20 and 40 percent in Mexican varieties; and Bursa has slightly less than 10 percent. Since the disease situation appears to be no better in Adana than in most other coastal provinces, it is interesting that the Mexican varieties are so much more successful in Adana: perhaps the reason for this is the use of a more complete package of new technology in Adana.

What is missing in Turkey for wider success with the Mexican wheats are replacement varieties that have high yield potential and adequate disease resistance. Up to now only the Mexican varieties that were originally imported have been grown in Turkey since their introduction in 1967; this has stagnated the production of the Mexican wheats. Of the total potential area in Turkey for fall planted, spring-type wheat, the Mexican wheats at present are covering approximately one-third and are accounting for about 15 percent of total wheat production, compared to more than 20 percent a few years ago. What is needed is more aggressive wheat breeding to develop varieties of the Mexican type specifically suited for Turkish growing conditions. The imported varieties were developed under Mexican conditions, in which there was no effective screening for stripe rust or septoria. Their susceptibility to these diseases makes wheat production more risky for the Turkish farmers.

The wheat production in the Aegean, Mediterranean, and Marmara regions accounts currently for nearly one-third of total wheat production in Turkey, compared with 25 percent before the introduction of the Mexican wheat technology. The production of Mexican wheats has annually contributed between 15 and 25 percent of the total wheat production in the country. If more resistant Mexican-type varieties or resistant high-yielding durum wheats are developed, the coastal areas' share of wheat production might increase further, since in the coastal areas about 1.5 million hectares of spring wheat are grown annually. The increased availability of irrigation water, land development, and improvements in water use efficiency might also have some favorable effects on the production of Mexican wheats. However, wheat production on fully irrigated land is relatively uncommon in Turkey and may dwindle in the future in competition with other crops, such as cotton, fruits, and vegetables.

In the irrigated coastal areas high-yielding American varieties of cotton, such as Delta Pine Smooth Leaf, have been introduced. With high cotton yields, better yields of wheat would be necessary to compete under the present price relationship.

Also there are different wheat cropping patterns in various areas of Turkey. Generally the Mexican wheats are grown only in the valleys or delta plains. The main irrigated valleys of the Aegean region have a typical distribution of crops according to the height of various terraces above the bottom of the valley. Usually only the bottom of the valley is irrigated. Some valleys have abundant water, while some are short of water. In the bottom of the valley cotton is grown on irrigated land, while tobacco, vegetables, and fruits are grown partly without irrigation. On the first terrace above the bottom of the valley, which is usually unirrigated, the products grown are tobacco, olives, grapes, figs, wheat, and barley. On the second terrace olives, grapes, wheat, and barley are grown. Then follow the slopes of the sides of the valley, where wheat or barley are grown.

Continuing programs for land leveling also influence the cropping pattern. When the farmers plant a wheat crop the land leveling is done during the summer season. Extensive leveling has recently temporarily increased the irrigated wheat area in the Adana area. When the leveling is completed the farmers most likely will concentrate on cotton. In Adana wheat is also a main crop in the unirrigated areas and in the foothills. In Anatolia cotton and citrus are dominating the cropping pattern more and more.

Small Farms Show Lower Wheat Yield

For the 1967-68 season the Ministry of Agriculture kept detailed records on 158 farms in the south Marmara region, and according to these records the small farms realized approximately 30 percent lower yields from the Mexican wheats than the large farms, because of a less complete use of improved practices. The use of drills for planting the seed was more common on the larger farms, and this factor alone accounted for two-thirds of the total difference. It seems also that credit was less readily available to operators and tenants of small farms.

In the Marmara region only about 10 percent of the farms participating in the program planted more than 50 hectares of wheat, and more than 80 percent had less than 10 hectares each. The average Mexican wheat area per farm recorded during the first year of the program was about 10 hectares. For later years there is no statistical information available, but most likely, as in other countries, the new technology has gradually sunk down to all sizes of farms in areas where it is suitable and profitable.

Since the Mexican wheat area in recent years has covered only approximately 7 percent of the wheat acreage in all of Turkey and 40 percent of the wheat acreage in the coastal areas the impact of the Mexican wheat on employment, income, and other sociological factors is in most cases likely to be limited. In spite of the relative high importance of wheat for the country as a whole, such other crops as cotton, fruits, and vegetables are more important to the gross production value in many of the coastal areas. Total additional annual wheat production resulting from the 600,000 hectares of Mexican wheats would mean a level of 1.2 million tons and a value of 1.2 billion TL (TL 1,000 per ton).

This additional income has undoubtedly resulted in an improvement in the level of family living for a large number of farm families. An increase in family income of about 50 percent has been likely for many farms. Part of it has been invested to improve the practices used in the production of wheat and other crops, including the use of combined seed-fertilizer drills and other forms of mechanization, land leveling, and improved irrigation methods. The areas where most of the Mexican wheats are grown, such as the Adana and Izmir areas, are also the areas where the interest in land leveling and improved irrigation is most pronounced. This of course does not mean that the farmers are improving the land mainly with the purpose of growing more wheat. Wheat might ultimately go, more or less, out of the cropping pattern for irrigated land in the coastal areas, and be more concentrated in the nonirrigated land in the valleys and the foothills.

MAIN FACTORS IN SUCCESS OF FIRST PHASE OF INTRODUCTION

The main reasons for the successful introduction of the Mexican wheats in Turkey can be summed up as follows:

1. Turkish agriculture was already at a rather sophisticated level in the coastal areas; many of the farmers were cotton and vegetable farmers, and they already had a fair degree of mechanization, used fertilizers, and so on.

2. The initial program was spearheaded by a group of progressive farmers who, along with the other farmers in the coastal areas, were willing to take a chance with the new wheats in the years that followed.

3. Without the determination and influence of the Turkish minister of agriculture, Bahri Dagdas, opposition to the program would have slowed it down, and possibly killed it before it got off the ground.

4. Credit was readily available to the farmers in such a form that a farmer who obtained the new seed also had to buy the recommended high doses of fertilizer (120 kilograms of nitrogen per hectare and 60 kilograms of phosphate per hectare).

5. The program was sold in the form of a complete package that included the use of combined seed-fertilizer drills, the practice of planting at the proper time and at the proper depth, and weed control, together with the new varieties and high rates of fertilizer application.

6. An extensive educational program was developed to inform the growers about the new varieties and about the cultural practices required to grow the seeds.

7. There were favorable price relationships that gave the farmers a fair return on the new wheat.

8. When adverse publicity was leveled at the Mexican wheat immediate steps were taken to instill confidence in the program.

The program differed from the earlier main programs for the introduction of Mexican wheats in India and Pakistan in the following ways:

1. The varieties and supporting technology were less adequately tested in Turkey by research (the testing that ultimately led to the large Turkish seed import was done by a group of progressive farmers) than in India and Pakistan.

2. Outside the province of Adana there had been practically no previous demonstrations of the new technology in farmers' fields, while in Pakistan and India quite extensive demonstrations had taken place before the main campaigns for introduction were launched.

3. The extension program in Turkey was relatively much stronger (250 extension agents, supported by 15 foreign extension experts well versed in the new wheat technology, covering 60,000 farmers and 170,000 hectares of wheat) than in India and Pakistan.

4. The educational program was carried out entirely through the extension service with the research service playing a rather insignificant role compared with similar programs in other countries.

5. There were no significant government subsidies for inputs involved, as there were, for example, in Pakistan.

The lessons that were learned from the experience with the wheat program in Turkey were the following:

1. Through a well-planned, strong educational program, rapid introduction through large seed imports is possible even without previous widespread demonstrations in the farmers' fields.

2. The experience in Turkey proves that the extension service can be very useful in the introduction process. It appears that the extension agents, if they are trained properly and provided with transportation facilities, can do an effective job.

3. Changes in the regulations for obtaining credit can be a valuable support for the program (this has been neglected in many other countries).

4. The Turkish experience underlines the importance of full support from a strong top man in the government, which also was found to be a main factor for success in Pakistan and India.

The successful introduction of 22,000 tons of wheat seed, of varieties which local wheat scientists felt had not been adequately tested and demonstrated, and which required a completely new set of cultural practices, is a dramatic story--one that proves how rapid introduction of new technology is indeed possible through a well-planned extension effort. In spite of the disease problems the crash program is proving to be a great blessing for Turkey, and is expected to have a lasting impact on wheat production in the coastal areas.

The Turkish success story confirms the following statement that was made by Borlaug: "The production of research information, data and materials does not necessarily mean an increase in food production. Someone must fit the research data into a meaningful plan to bring them to bear on food production problems, before human needs are met" (Borlaug 1968).

While the Turkish program differed from the earlier programs for the introduction of the Mexican wheats, it emphasizes, more strongly than most of the other programs what can be achieved through the so-called "kickoff approach" widely used for introduction of the Mexican wheats (Borlaug and Aresvik, 1973): the introduction of improved technology through strong national production campaigns for one main product at a time, as new technology becomes available and is tested.

The experience in Turkey can be very useful to other nations in planning "wars against hunger" in areas in which well-tested and improved technology is available.

It will always be a question of judgment whether improved technology is well tested and adapted. Some scientists seem to prefer to go on testing for a long period of time. However, no war, and least of all the war against hunger and famine, can be won without taking calculated risks. Politicians such as Minister Dagdas often seem to be in a better position than scientists and planners to determine the feasibility of a campaign for the introduction of improved technology. Also they are the ones who ultimately decide and who have to take the final responsibility for the decision. In spite of certain difficulties and setbacks, the recent events of the Green Revolution seem to indicate that a certain boldness pays off handsomely.

CHAPTER

13

INTRODUCTION OF IMPROVED TECHNOLOGY IN THE WINTER WHEAT AREAS

The introduction in Turkey of the program for developing new high-yielding winter wheat varieties, which has been referred to as the "High-Yielding Winter Wheat Project," coincided with the program for the introduction of the Mexican spring-type varieties; but the winter wheat program was much more modest and was developed more gradually. The large seed import carried out by Turkey in 1967 included 400 tons of U.S. varieties of winter wheats. These winter wheats were planted in the fall of 1967 on about 4,800 hectares of state farm land, with deep furrow drills, also imported from the United States. In the fall of 1968, 38,000 hectares were planted with the U.S. varieties.

In the fall of 1967 Turkey also imported 100 tons of Bezostaya, a high-yielding Russian winter wheat. This wheat had been tested in Turkey two years before and had shown a good adaptation potential. The 100-ton acquisition was negotiated when Suleyman Demirel, who was then prime minister, made an official visit to Russia. In the fall of 1968 about 7,200 hectares of Bezostaya were planted in Turkey.

The average winter wheat yield in the 1969 Turkish harvest was 1.9 tons per hectare for the American varieties--mainly a variety called Wanser--and 2.4 tons per hectare for Bezostaya, as compared with 1.5 tons per hectare for domestic varieties.

In the fall of 1969 about 68,400 hectares of high-yielding winter wheat were planted, with less than 4,000 hectares devoted to Wanser, and the rest of the acreage going to Bezostaya. The highly adaptive Bezostaya has stiff straw, and is resistant to stripe and leaf rusts, but is susceptible to stem rust. The major objection to Bezostaya is its red grain. The Turks prefer white grain since they generally grind the entire kernel for their bread. The general bread-baking quality of Bezostaya is good.

In February of 1970 the Cereals Production Branch of AID presented a plan to the Turkish Ministry of Agriculture for a controlled program aimed at increasing winter wheat yields by an additional one ton per hectare on one million hectares of land in selected provinces of the Anatolian plateau and Thrace. The program was designed to use new high-yielding winter wheat varieties, especially Bezostaya and Wanser, together with recommended fertilizer rates and cultural practices. The Ministry of Agriculture responded well to the plan, but requested that the program also include three high-yielding Turkish winter wheats that had been recently introduced. The result was that in the fall of 1970 about 287,000 hectares were planted with high-yielding imported winter wheats, including about 250,000 hectares of Bezostaya and 37,000 hectares of Wanser, and additional areas were planted with improved Turkish varieties; thus a total area of approximately 1.0 million hectares was devoted to improved winter wheat varieties. Most of the seed exchanged hands between farmers, three times as much as government grown seed. (See Table 13.1.)

Before the wheats were planted an educational program was carried out through regional meetings for county agents and the preparation of brochures and posters for the farmers informing them about the new varieties and the fertilizer rates recommended. Some 15,000 copies of each of two posters and 70,000 copies of each of two brochures were printed and distributed for the program.

With the help of this program, it is estimated that on the one million hectares of land covered by the program at least 0.75 million tons of additional wheat was produced, with a gross value of TL 750 million. It was concluded that the use of more fertilizer could have produced considerably better results.*

For the 1971-72 season an estimated total area of 400,000-500,000 hectares of Bezostaya and Wanser (predominantly Bezostaya) was planted. Seven provinces were

---

*Harvey P. H. Johnson, "High Yielding Winter Wheat for Turkey: A Progress Report," AID: Turkey, August 1971. (Mimeograph)

TABLE 13.1

Hectares of Wheat Planted, by Variety, in 15 Turkish Provinces During the High-Yielding Winter Wheat Project, 1970-71

	Hectares							Tons of Seed Used in 1970-71 Crop	
Province	Bezostaya	220/39	Wanser	1593	4/11	Total	Project Goal	Grown by the Government	Grown by Farmers
Konya	64,672	169,200	12,488	61,408	29,562	337,330	330,000	3,527	38,073
Afyon	7,012	10,000	3,110	n.s.	416	20,538	70,000	885	2,334
Corum	4,634	5,093	1,161	734	n.s.	11,622	56,000	797	863
Yozgot	19,350	55,755	n.s.	1,214	n.s.	76,319	90,000	1,505	9,485
Sivas	2,786	46,529	n.s.	1,000	n.s.	50,315	80,000	1,257	5,788
Kayseri	376	65,438	2,507	764	n.s.	69,085	70,000	2,965	7,243
Nigde	6,380	70,154	3,465	1,610	n.s.	81,609	45,000	2,628	9,036
Nevsehir	3,020	44,257	1,050	300	n.s.	48,627	45,000	2,400	4,504
Kirsehir	3,402	29,036	714	3,535	280	36,967	50,000	1,141	4,140
Eskisehir	6,950	13,820	1,338	985	7,535	30,628	40,000	2,225	1,907
Ankara	38,860	110,163	11,300	24,600	5,040	189,963	270,000	9,374	18,956
Edirne	23,425	n.s.	n.s.	n.s.	n.s.	23,425	30,000	1,305	2,041
Tekirdag	41,250	n.s.	n.s.	n.s.	n.s.	41,250	45,000	1,768	4,125
Kirklarili	26,415	n.s.	n.s.	n.s.	n.s.	26,415	24,000	1,057	2,717
Cankiri	1,992	2,200	n.s.	n.s.	n.s.	4,192	25,000	288	312
Total	250,524	621,645	37,133	96,150	42,833	1,048,285	1,270,000	33,122	111,524

Notes: Bezostaya is a Russian variety, Wanser is American, and the other three varieties listed are Turkish.

n.s. = Not significant amounts.

Source: Food and Agriculture Division, AID: Turkey.

added to the 15 covered by the 1970-71 program. Bezostaya has done well in Thrace, as a result of relatively widespread rainfall, and estimates show that nearly 80 percent of the total wheat area in Thrace might have been planted with Bezostaya in the fall of 1972. On the Anatolian plateau only an estimated 5 percent of the total wheat area might have been planted with Bezostaya or Wanser, and a widespread introduction of foreign high-yielding winter varieties is not yet in sight. The main reason for this is that generally winter wheat yields in the central plateau are not limited by the capabilities of the local varieties that have been grown in the area. Also the improved Turkish varieties generally yield much less than their potential. According to research data reported by the Wheat Research and Training Project, the response to the use of nitrogen fertilizer has not been particularly strong in central Anatolia, while there appears to be a generally wide response to the use of phosphorus up to a level of 40 kilograms per hectare. The relatively weak response to nitrogen is caused by other factors that limit yields to a low level, mainly a shortage of water, late planting, and unsatisfactory seed bed preparation. Under the traditional wheat-fallow-wheat rotation the farmers wait for the first rain before preparing the seed bed. The seed does not germinate until the second rain, which is usually too late for the establishment of a good stand of wheat before the cold weather sets in. Under such conditions only relatively limited results can be achieved through a change of seed variety combined with the application of fertilizer. After the fallow year there is usually already enough nitrogen in the soil (up to 70 kilograms of nitrogen have been found down to a depth of 1.5 meters) to achieve the yield level as determined by the other limiting factors. Consequently a high-yielding variety cannot show its potential even by the addition of nitrogen.

In the opinion of wheat scientists the factor that most seriously limits wheat production in the dry areas of Turkey is the management of the soil during the fallow period and the subsequent management of the wheat crop. This was recognized at an early stage, since the traditional summer fallow tillage method produces relatively low unit yields and sharp variations in annual production. Under the AID-Oregon State University cereals project, which initially emphasized the introduction of the high-yielding Mexican wheats, research was started in 1969 to test the successful summer fallow techniques practiced in the northwestern region of the United States on three Government of Turkey state farms and one research station in Turkey. The wheat producing areas in the northwestern region of the United States have climatic conditions that are rather similar to the dry-land wheat areas of Turkey. Later the Government of Turkey invited the Rockefeller Foundation to assist in the formation of a nationwide, interdisciplinary research organization in Turkey that would aim to raise wheat production and stabilize it at a higher level. Scientists representing both the Rockefeller Foundation and the AID-Oregon State University cereals project are now working closely with Turkish scientists to solve the wheat management problem in Turkey. However, it takes two years to complete a crop cycle, and because weather influences wheat management practices to a large extent, and weather patterns vary widely from year to year, several years would be required to perfect a workable management system. Early efforts in this direction have proven encouraging, and a large-scale introduction of high-yielding wheat technology in the central plateau will be possible as soon as an improved management system has been more fully developed. The main problem concerns not seed varieties, but production management and moisture conservation.

For Turkey as a whole the achievements in the field of wheat production have already been substantial. The introduction of the Mexican varieties, relatively high-yielding Italian durums in many of the durum-growing areas, and high-yielding winter wheat varieties in certain areas, together with the use of additional fertilizer and improved practices, has resulted in an estimated 2.0 million tons of additional wheat amounting to a value of TL 2.0 billion.

This estimate is based upon the calculated increases resulting from the introduction of both high-yielding spring wheats and winter wheats, and is supported by an estimate of increased use of fertilizer on wheat. The increase in fertilizer used for wheat after the 1966-67 season represents approximately 169,000 tons of nutrient. If multiplied by a factor of 12, which seems reasonable, this would give about 2.0 million tons of wheat. Turkish wheat production for 1967 was approximately 9.8 million tons: if this amount can be regarded as a normal year's production, prior to the introduction of new wheat technology, a normal year's production at present would be of the level of 11.8 million tons, or possibly 12 million tons.

Through the strengthening and reorganization of wheat research activities that have taken place, and the continued support of AID, Oregon State University, and the Rockefeller Foundation, the prospects are bright for future increases in wheat production in Turkey--possibly a doubling or even a tripling of the average wheat yields in the various areas compared with the yield level before the introduction of new technology.

CHAPTER

14

SUMMARY AND CONCLUSIONS

The republic of Turkey gained national independence in 1923 under the leadership of Mustafa Kemal (Ataturk), who was elected the first president of Turkey. The new nation, based on the principles of nationalism and secularism, was firmly committed to the achievement of modern Western civilization and to the establishment of democracy.

In the years that have followed a political and social transformation has taken place in Turkey, and in spite of the presence of important difficulties, political stability and social tranquility have prevailed. In the major areas of daily living the Westernization of Turkey is now complete and irreversible, the principle of democracy is firmly established, and the technological and economic development is considered successful, by any standard.

The total population of Turkey has more than tripled since 1923, and is currently estimated at approximately 36 million, of which 60 percent is rural. Per capita real income, increasing steadily, has now reached the equivalent of about $450 U.S. per year: this is almost four times the per capita income in India, and approximately one-sixth that of Northwestern Europe. Agriculture continues to be the mainstay of the Turkish economy, although its share of national income has declined from more than two-thirds to one-fourth. The population growth rate of Turkey is about 2.9 percent, with the recent rural population growth estimated at about 1.5 percent. There is still a serious problem of unemployment and underemployment in Turkey. Nevertheless, Turkey has made significant economic and social progress especially since the end of the Second World War. While the share of agricultural production in GNP has fallen from 45 percent to 26 percent, the share of industrial production has increased from 13 to 18 percent. The network of all-weather roads has been doubled. The literacy rate has increased from about 30 percent to about 50 percent. The number of students in primary schools has increased by over 65 percent.

Turkey's agricultural development over the past 50 years has been adequate to produce most of the foods and fibers needed for her growing population, and agricultural products in recent decades have accounted for about 80 percent of total Turkish exports. The area cultivated has more than tripled, keeping the area cultivated per capita approximately constant. The grazing area, however, has decreased to less than one-half. The areas devoted to fruits and vegetables, vineyards, and olive groves have nearly tripled, with the major portion of the increase taking place over the last 20 years. The area of agricultural land under irrigation has increased considerably, especially in the coastal regions, facilitating a more intensive production of high-value crops. The production of industrial crops has increased rapidly. There are, however, significant regional differences, with the most rapid crop development occurring in the coastal regions. The total area of arable land and land devoted to permanent crops is approximately 27.4 million hectares. This represents 35.5 percent of the total Turkish land area and cannot be significantly increased.

Most of the people in the agricultural areas live on small family-owned farms. There is a total of about 3.1 million farms, giving an average farm size of nearly 9 hectares. The traditional Turkish rules of inheritance are causing the subdivision of large farms to proceed rapidly. The relative importance of small holdings has been increasing; while in 1952 holdings with less than 5 hectares of arable land cultivated 19 percent of the total, these holdings cultivated 31 percent in 1971. Approximately 84 percent of the farmers are cultivating their own land. In addition 12 percent own a part of the land that they cultivate. The situation regarding land fragmentation has improved considerably during the last two decades, with the percentage of single-unit holdings increasing from 5 percent in 1950 to 22 percent in 1970.

The relative importance of plant production in Turkey has been increasing: it represented 54 percent of agricultural production in 1938 and nearly 68 percent in 1970, while the share of livestock production in the same period declined from 39 percent to 27 percent. The share of fruits and vegetables increased from 11 percent in 1952 to 21 percent in 1970. In crop production the grains have been dominant, accounting for 85 percent of the sown area, with wheat alone in 1971 having a share of 55 percent. By comparison the share of wheat was 45 percent in 1951. In production value its share in 1971 was 26 percent and had decreased 20 percent compared with 1951. The industrial crops have, on the other hand, increased their share of production value and are approaching 20 percent. This development has taken place, in spite of a decreasing growing area, as a result of rising yields and profitability.

The following growth patterns have occurred in Turkish agriculture:

1. Modernization has developed first in the coastal areas, which have relatively widespread rainfall, supplementary irrigation, and a mild climate.
2. The development has mainly taken place through the introduction of new improved technology, such as that represented by the new sugar beet crop in central Anatolia, the tea and tangerine orchards on the coast of the Black Sea, the citrus and banana crops and truck gardens on the southern coast, or the new high-yielding varieties of cotton and wheat, combined with increased fertilizer applications, more and improved irrigation, mechanization, and plant protection chemicals.
3. Large and medium-sized farms have led the way in the modernization process.
4. This process has progressed mainly in plant production, while animal production has remained more stagnant because of a limited introduction of new technology.
5. The differences in modernization between various regions and groups of farmers in Turkey are strongly influenced by the risk factor. The risk and uncertainty in agricultural production is much higher in the dry, rainfed areas than in the wetter, highly irrigated coastal regions. Moreover small farmers operating at marginal living standards cannot take the economic risks that are involved in innovations.
6. The rapid introduction of Mexican wheat varieties in the coastal areas represents one of the best examples of how improved high-yielding technology can be successfully introduced through a strong, well-planned effort.

The main factors that have determined the growth patterns are the following:

1. The cultivated area increased from 7.6 million hectares in 1927 to 24.4 million hectares in 1971.

2. The expansion of the cultivated area has led to a reduction in the grazing area over the same period, from 37 million hectares to 19 million hectares; some land has been rendered unfit for production, as a result of the soil erosion caused by expansions into submarginal areas.

3. There have been pronounced increases in yields for most crops: from the 1946-50 period to the 1966-70 period wheat and barley yields grew about 2.0 percent annually, and yields of cotton and sugar beets rose 4.6 percent and 3.4 percent, respectively.

4. Through investments in dams an increasing quantity of irrigation water has become available, and in spite of relatively inefficient use, irrigation has contributed greatly to increased cropping intensity and yields.

5. There has been a rapid increase in the use of fertilizer: for the 1961-71 period the annual compound rate of total increase was 29 percent, and for fertilizer used on wheat it was nearly 40 percent.

6. The effects of increased fertilizer use have been compounded through the introduction of high-yielding fertilizer-responsive varieties of wheat, cotton, and sugar beets.

7. Rapid mechanization through the use of tractors has taken place since 1948, with an annual compound growth rate in tractor use of 20 percent; the initial burst in tractor mechanization contributed more to a rapid expansion of the cultivated area than to increased yields.

8. There has been a considerable increase in the amount of organized credit available through the Agricultural Bank and the cooperatives supervised by the bank; however, the amount of credit is still highly inadequate, especially for the small farmers.

9. The price relationships for various products have been relatively favorable for agricultural development, as a result of growing domestic and external demand, government price supports that have achieved price stabilization for many products, and decreasing real prices for fertilizers.

10. The government's agricultural research and field services have been strengthened, although these services are still inadequate.

11. The political leadership in Turkey has been able to maintain peace and political stability since the establishment of the new republic, and this has favored a generally uninterrupted flow of economic and social development.

12. Turkey's proximity to the European market and her association with EEC since 1963 have been beneficial to her agricultural development. Europe is responsible for about 75 percent of both Turkish imports and exports.

13. The construction of Turkey's all-weather national highway network has exerted a profound influence on her agricultural development; the construction of feeder roads, however, was delayed to concentrate on highway construction.

14. The improvements in the field of general education, the rapid increase in the availability of transistor radios, and improvements in transportation and communications have facilitated the spreading of new ideas and of improved agricultural technology.

15. The government, especially when the Republican party has been in power, has favored industry at the expense of agriculture, but part of the industrial development has directly and indirectly facilitated agricultural development.

16. In spite of the lack of effective agrarian reform, some improvements in the agricultural structure have taken place: the fragmentation of farms has been reduced, many large farms have been split up, and the high degree of self-ownership has been maintained.

In spite of this progress Turkish agriculture faces the following major problems noted in this book:

1. There is persistent poverty among a large proportion of the rural population. On the average the per capita income in agriculture has been about half of the average per capita income in Turkey and below one-fourth of the income in Turkish industry. Various studies (cited in this book) of rural income distribution in Turkey find that the income for the majority of the farm families is below the poverty level, that there is a great disparity compared with the nonagricultural sectors, that the income distribution within the agricultural sector is rather unequal compared with most developed countries, and that there are large farm income fluctuations over time.

2. There is a lot of disguised seasonal unemployment as well as underemployment in agriculture. In the Second

Five-Year Plan seasonal unemployment in agriculture is estimated at 910,000 persons. Because agriculture is already overpopulated and because of the high birth rate among farm families, unemployment is a very difficult problem to deal with. The ultimate solution would involve the transfer of farm population to other sectors, as well as training and relocation within the farm sector.

3. Turkish agricultural development has been geographically uneven. Modern commercial agriculture is practiced mainly in Cukurova and Antalya in the south, along the river valleys lying toward the Aegean Sea in the west, and on the plains south of the Marmara Sea. The rapid progress that has taken place in the agriculture of these regions has not been matched in the plateau areas, and the least developed agriculture is found in the eastern provinces, where 83 percent of the population is engaged in agriculture, low levels of output and welfare are maintained, and primitive pastoral farming is heavily relied upon.

4. There has also been a pronounced imbalance among the various agricultural subsectors. Rapid development and modernization have taken place mainly in crop production, while the livestock sector still emphasizes traditional practices, and has a low growth rate compared with the growth in demand for livestock products.

5. In spite of a relatively high agricultural growth rate in Turkey over an extended period of time, the growth rate of demand for many products, particularly for livestock products, has been higher because of a high population increase and rising income per capita. This has resulted in growing deficits and rising real prices for livestock products. Only a strong, effective effort to modernize and increase livestock production can change this trend.

6. The high rural birth rate, especially in eastern Turkey, has impeded agricultural development and an increase in rural per capita income. One consequence of this is the annual influx of about 200,000 villagers into the cities, with many villagers also going abroad to find employment. Another result is that concomitant with the increase in absolute numbers of rural inhabitants (1.5 percent) and with increasing tractor mechanization, the number of landless workers is growing, thereby adding to underemployment and other sociological problems. The family planning program adopted by Turkey in 1965 needs to be more effectively implemented in the rural areas.

7. There are growing urban-rural disparities: a gap between the educated urban elite and the uneducated, rural masses inhibits overall development. City-bred, city-trained agricultural extension workers and other members of the educated elite who work in the rural areas generally cannot communicate well with the farmers and tend to talk down to them. The traditional cultural disparity is being intensified by growing income disparities between city workers and farm workers.

8. The Turkish bureaucracy is both a source of strength and a constraint on overall development, and this may also hold true for agricultural development. Bureaucratic rules made it difficult, for example, for Turkey to import improved cereal varieties, such as the Mexican wheats, until the minister of agriculture became personally involved in the program.

9. Turkey's achievement of full membership status in EEC will open the Turkish economy to foreign competition, which in turn will open up new avenues for modernizing commercial agriculture in Turkey. At the same time this will call for high productivity and high quality products together with low costs of production and prices that are competitive without having to be supported by subsidies: it represents both an opportunity and a challenge.

The following is a summation of the impediments to a rapid increase in Turkish agricultural productivity together with suggestions for alleviating and eventually removing these impediments:

1. The introduction of improved technology to the more than 3.0 million farmers living in 40,000-50,000 villages and other settlements in Turkey represents a formidable task. In areas with tightly knit village communities the farming system ties the villagers together; thus a new crop rotation might be difficult to introduce to individual farmers. The best strategy might be to work with the village as a unit.

2. Turkish farmers are strong, hardworking people who are anxious to improve their lot and are not against innovation; however, the small, marginal farmers have limited resources for taking risks. It is therefore important to adopt packages of improved high-yielding technology that carry a small risk in relation to expected benefits. The possibilities of crop insurance plans might be explored, and land utilization systems might be

established that would guarantee that the farmer reaps the benefits of the innovations he adopts. Also the Turkish credit institutions must adopt a flexible policy in cases of crop failure.

3. The majority of small farmers lack adequate general and technical education. A recent study by the State Planning Organization indicates that only 39 percent of Turkish villagers over 16 years of age claim literacy,and only 10 percent of villagers over 16 claim to have graduated from primary school (the fifth grade). Compulsory primary education is still not enforced in 15,000 villages, and 5,000 villages or more are so small and isolated that they cannot justify building a school and employing a teacher. In the eastern, and especially the southeastern, provinces, which are overwhelmingly rural, school attendance and graduation levels are well below the levels in western Turkey. In addition there are limited facilities for technical agricultural training for prospective farmers. Provision of both general and technical education facilities for the village population would greatly facilitate the introduction of modern agriculture.

4. About one-third of the villages are located in mountain and forest regions that lack the necessary transportation and communication facilities. The real costs of building and maintaining the necessary feeder roads in these areas would be limited if labor-intensive methods were used in areas with widespread seasonal unemployment.

5. The agricultural credit situation in Turkey is unsatisfactory: it is estimated that about 40 percent of the farms have no access to credit, and that almost half of those that have access to credit can secure only insufficient loans. The introduction of improved agricultural technology requires capital, and reorganization and expansion of Turkish agricultural credit facilities is therefore essential to achieve more rapid agricultural development.

6. There are broad deficiencies and shortcomings in the marketing system, especially for supplying current export markets and for accomplishing a profitable integration with EEC through the production of high-quality products, improved grades, product standardization, and effective packaging. Similarly the supply mechanisms for farm inputs must be improved in order to provide to all farmers the various inputs required (at specified periods) for a more productive technology.

7. In some parts of Turkey the structure of agriculture--the land tenure situation--might represent an

impediment. However, the inequality in land ownership is much less pronounced than in many other developing countries. In some areas of Turkey, notably in the southeast, absentee landlords are common, and in many areas the political power of major landowners might result in political and social unrest if the inequalities in ownership are not reduced through the redistribution of land. On the other hand the government must be careful not to reduce output and slow down progress by breaking up a large number of productive units.

8. An effective administrative reorganization of the Ministry of Agriculture and other Turkish ministries and agencies serving agriculture seems to be necessary. In Turkey there is currently no single public authority responsible for agricultural development and policy. The Ministry of Agriculture has almost no involvement with the economic aspects of agriculture; it is primarily involved with the technical aspects. The present situation impedes agricultural development by creating overlapping services and unnecessary red tape and bureaucracy. It would be more efficient to give the Ministry of Agriculture full command of Turkey's agricultural development policies by having it coordinate all public services related to agriculture.

9. Rapid introduction of modern high-yielding technology among the farmers may require a slightly different type of agricultural student than Turkey has had previously. The majority of the agricultural students have been city boys without solid practical experience in agriculture, and their education has emphasized book learning. In addition to attempting to raise the academic standard, which is already going on, it would be important to give preference in the agricultural schools to village boys who have practical experience, and to provide the students with better practical training in modern technology.

10. A traditional type of agriculture cannot be transformed into a more modernized and productive agriculture without the introduction of improved technology. A number of studies have shown that investment in agricultural research can be highly profitable, and usually buys more growth per dollar than other forms of investment in agricultural development. The modernization of Turkish agriculture will require a much more solid research underpinning than is found in Turkey at present. Many of the more important agricultural problems do not seem to have received the attention that they have deserved. For example, the introduction of improved cotton and wheat

varieties was not spearheaded by research scientists, but by innovative farmers. This would seem to call for reorganization and streamlining of the Turkish agricultural research service aimed at maximum efficiency and considerable expansion.

11. To achieve more rapid introduction of improved high-yielding technology, more effective agricultural extension will also be needed. The present extension staff is overburdened with reporting and regulatory functions, and often the agents lack the practical experience and farming background to communicate successfully with the farmers. Also the introduction of profitable packages of improved technology has often been missing because of an inadequate research underpinning and a lack of proper contact with the research service. Effective in-service training programs in improved farming technology would help to better this situation.

12. Because of the relatively limited arable land resources in Turkey, improved technology can be introduced (as was done for the Mexican wheats) by staging strong national or regional production campaigns for one key crop at a time, using the so-called kickoff approach. To spur production such campaigns could be developed as new technology becomes available and is tested.

13. The generally low level of water use efficiency in Turkey represents an impediment to agricultural development in the irrigated areas. Proper land leveling is essential for efficient water management. The returns on an effective program for improved water management could be very high, and such a program deserves top priority for the irrigated areas. Technical assistance to the farmers and demonstrations of the benefits of such a program would be important to encourage widespread adoption of improved water use practices. The farmer-financed, private contractor-executed, on-farm development program maintained in the Izmir area seems to offer the best prospects for rapid improvement of the water use situation, which is urgently needed for most of the irrigated land.

14. Economic planning is relatively well developed in Turkey. However, in spite of improvements over the last decade, agricultural planning is still inadequate to achieve rapid agricultural development. The Planning and Economic Research Department of the Ministry of Agriculture could be further strengthened, and the program for the introduction of improved technology could be planned and executed in closer cooperation with technical specialists and agencies.

15. Finally, in order to ensure more rapid agricultural development, higher priority could be given to agriculture, not only in terms of policy, but also in the allocation of increased resources. Even during the period of the five-year plans agriculture has not been given the priority which it has deserved, and which is needed for more rapid development and balanced economic growth. The rapid growth in Turkey of both population and food demand, and Turkey's full integration into EEC will make it more urgent than ever before that agriculture be given the proper priority.

POTENTIALS AND PROSPECTS FOR FUTURE DEVELOPMENT

As indicated throughout this book, Turkey has a great potential for agricultural development. The country has considerable water resources that can be further developed to widely expand the irrigated area. The cropping intensity can be increased through multiple cropping in the already productive irrigated areas, through a reduction in fallow land, and through improved water conservation practices in the rainfed areas. The level of crop yields in many irrigated areas can be doubled, or increased even more, on the basis of current technological development; future progress in agricultural technology will make further yield increases possible. The overall technological advances and increased crop yields will lead to large increases in total production and still greater increases in the total value of production, with a higher percentage of the arable land being used for high-value crops.

In the Second Five-Year Plan it is indicated that the potential exists in Turkey to irrigate 7.0-8.5 million hectares of land, compared with the 1.5 million hectares irrigated at present. This would mean that the present irrigated area would more than quintuple, and that the percentage of arable land and of permanent irrigated crops would increase from nearly 6 percent (at present) to nearly 30.0 percent. Through improved land and water management and the introduction of high-yielding technology the levels of cropping intensity and crop yields could reach those levels attained in the most agriculturally developed countries. This achievement alone could make Turkey an agricultural giant.

Through a much stronger research effort crop yields and cropping intensity could also be considerably increased on the rainfed land. The continuing research designed to increase wheat yields through improved cultivation practices, which permit both the early establishment of the wheat seedling and the greatest conservation of rain falling on the fallow land, will show the way. Improvement of the fallow system in combination with high-yielding varieties and other improvements might make doubled crop yields possible. However, the improved cultural practices will not be adapted by a livestock-oriented economy that depends on stubble and weeds that come up in summer fallow as one of the main sources of feed. Concomitantly a strong research effort could also be initiated for testing the crop rotation system used in areas of Australia that have a Mediterranean-type climate--that of growing fodder legumes in alternate years in rotation with wheat. These legumes have a fairly high percentage of hard seed that remain ungerminated in the wheat cropping year, thus providing self-perpetuating pastures. This system improves the soil structure and provides free nitrogen for the wheat crop. It is possible that the wheat-legumes rotation system can be more economical in some ecological areas of Turkey than an improved fallow system, especially with the higher prices of nitrogen fertilizer likely to prevail in the future.

In order to fill the technological gap between the improved technology currently applied in the most developed countries and the present technological level in Turkey a generally much stronger and more efficient research effort is needed. This would require a streamlining of the agricultural research organization and an overall greater emphasis on agricultural research. There are strong indications that Turkey typifies the case of a country with considerable underinvestment in agricultural research. The total area of arable land and land devoted to permanent crops cannot be significantly increased; thus progress must come through raising the level of agricultural productivity. In this way a balanced economic development can take place in Turkey. The country can achieve self-sufficiency in food and fibers, plus considerable exports of fruits and vegetables, even with a population that is double the present figure. With the present rapid population growth in Turkey, the population may double in about 25 years. Because the arable land area and other resources are limited, this rapid population growth cannot continue without very serious long-run consequences, in the form of poverty, overcrowded cities, and food supply problems. The good potentials and prospects for further agricultural development in Turkey should not in any way be used as an excuse for the country's present inadequate effort to introduce family planning. The welfare of future generations in Turkey can be secured only through family planning programs that successfuly reduce the high rate of population growth.

STATISTICAL APPENDIX

TABLE A.1

Data Relating Agriculture to the Rest of the Turkish Economy

Year	Population (million)	Rural Population (million)	Gross National Product (TL billion)*	Agricultural Product (TL billion)*	Total Exports ($ million)	Agricultural Exports ($ million)
1961	28.5	19.2	49.1	17.3	347	274.1
1962	29.2	19.5	52.1	18.3	381	296.1
1963	29.9	19.9	56.1	19.7	368	284.3
1964	30.6	20.2	58.9	19.7	411	344.9
1965	31.4	20.6	61.6	19.0	464	398.1
1966	32.2	20.9	67.9	21.2	490	417.9
1967	32.7	21.1	72.1	21.4	523	463.9
1968	33.5	21.3	76.9	21.8	496	431.6
1969	34.7	21.6	81.7	21.8	537	442.1
1970	35.2	21.8	86.4	22.0	588	472.3
1971	35.7	21.9	94.3	24.1	677	529.8

(Based on 1961 prices.

Source: AID, Ankara, Economic and Social Indicators--Turkey, August 1972.

TABLE A.2

Data from Table A.1 Indexed on 1961 Basis

Year	Population	Rural Population	Gross National Product	Agricultural Product	Total Exports	Agricultural Exports
1961	100	100	100	100	100	100
1962	102	102	106	106	110	108
1963	105	104	114	114	106	104
1964	107	105	120	114	118	126
1965	110	107	125	110	134	145
1966	113	109	138	123	141	152
1967	115	110	147	124	151	169
1968	118	111	157	126	143	157
1969	122	113	166	126	155	161
1970	124	114	176	127	169	172
1971	125	114	192	139	195	193

Source: Calculated by the author.

TABLE A.3

Data from Table A.1 Showing Percentages Accounted for by Agriculture in Turkey

Year	Rural population as Percent of Total Population	Agricultural Product as Percent of Gross National Product	Agricultural Exports as Percent of Total Exports
1961	67	35	79
1962	67	35	78
1963	67	35	77
1964	66	33	84
1965	66	31	86
1966	65	31	85
1967	65	30	89
1968	64	28	87
1969	62	27	82
1970	62	25	80
1971	61	26	78

Source: Calculated by the author.

TABLE A.4

Wheat and Five Other Leading Turkish Crops Ranked According to Area Figures for 1961 (thousand hectares)

Year	Wheat	Barley	Grapes	Corn	Cotton	Tobacco
1961	7,717	2,786	750	705	649	141
1962	7,800	2,800	773	667	660	149
1963	7,850	2,850	794	670	628	236
1964	7,870	2,750	800	680	680	272
1965	7,900	2,770	800	650	685	222
1966	7,950	2,710	830	655	712	285
1967	8,000	2,725	840	675	718	295
1968	8,250	2,730	848	655	713	273
1969	8,660	2,687	838	659	639	372
1970	8,600	2,590	845	650	528	328
1971	9,000	2,750	845	660	670	310

Source: State Institute of Statistics, Turkey, Summary of Agricultural Statistics, 1962-1972.

TABLE A.5

Percent of Total Area Sown in Each of Six Leading Turkish Crops in Table A.4

Year	Total Cropped Area (thousand hectares)	Area Sown Each Year (thousand hectares)	Fallow Land (thousand hectares)	Wheat (percent)	Barley (percent)	Grapes (percent)	Corn (percent)	Cotton (percent)	Tobacco (percent)
1961	23,076	15,128	7,948	51.0	18.4	5.0	4.7	4.3	0.9
1962	23,215	15,167	8,093	51.4	18.5	5.1	4.4	4.4	1.0
1963	23,913	15,276	8,547	51.4	18.7	5.2	4.4	4.1	1.5
1964	23,843	15,367	8,476	51.2	17.9	5.2	4.4	4.4	1.8
1965	23,841	15,294	8,262	51.7	18.1	5.2	4.3	4.5	1.5
1966	23,982	15,454	8,528	51.4	17.5	5.4	4.2	4.6	1.8
1967	23,836	15,513	8,383	51.6	17.6	5.4	4.4	4.6	1.9
1968	24,092	15,400	8,692	53.6	17.7	5.5	4.3	4.6	1.8
1969	29,672	15,848	8,824	54.6	17.0	5.3	4.2	4.0	2.3
1970	24,294	15,589	8,705	55.2	16.6	5.3	4.2	3.4	2.1
1971	24,424	15,821	8,603	54.4	16.3	5.3	4.0	4.3	2.1

Source: AID: Ankara, Economic and Social Indicators--Turkey, August 1972; and State Institute of Statistics, Turkey, Summary of Agricultural Statistics, 1971.

TABLE A.6

Wheat and Five Other Leading Turkish Crops
in Value of Production
(TL million)

Year	Value of All Crops	Wheat	Barley	Maize	Tobacco	Sugar Beets	Cotton Lint
1961	16,400	5,187	1,580	615	758	482	1,095
1962	17,500	7,071	2,065	544	962	325	1,284
1963	19,300	8,210	2,521	718	1,073	311	1,316
1964	19,100	6,756	1,875	695	1,768	486	1,706
1965	18,500	7,293	2,148	665	1,070	692	1,679
1966	20,700	8,630	2,709	799	1,327	499	1,955
1967	21,800	8,990	2,687	841	1,555	677	2,152
1968	22,200	8,730	2,642	830	1,393	767	2,400
1969	22,900	10,206	3,007	894	1,005	693	2,258
1970	23,400	10,192	2,256	990	1,048	671	2,884
1971	24,900	13,255	3,472	1,021	1,098	851	3,857

Sources: FAO, Production Yearbook, 1961-71; State Institute of Statistics, Turkey, Agricultural Structure and Production, 1961-71.

TABLE A.7

Percent of Total Crop Value for Each Crop in Table A.6

Year	Percent in Wheat	Percent in Barley	Percent in Maize	Percent in Tobacco	Percent in Sugar Beets	Percent in Cotton Lint
1961	31.6	9.6	3.8	4.6	2.9	6.7
1962	40.4	11.8	3.1	5.5	1.9	7.3
1963	42.5	13.1	3.7	5.6	1.6	6.8
1964	35.4	9.8	3.6	9.3	2.5	8.9
1965	39.4	11.6	3.6	5.8	3.7	9.1
1966	41.7	13.1	3.9	6.4	2.4	9.4
1967	41.2	12.3	3.9	7.1	3.1	9.9
1968	39.3	11.9	3.7	6.3	3.5	10.3
1969	44.6	13.1	3.9	4.4	3.0	9.9
1970	43.6	9.6	4.2	4.5	2.9	12.3
1971	53.2	13.9	4.1	4.4	3.4	15.5

Source: Calculated by the author.

TABLE A.8

Wheat Area in the Nine Agricultural Regions of Turkey
(thousand hectares)

Year	Region I (Central North)	Region II (Aegean)	Region III (Marmara)	Region IV (Mediterranean)	Region V (Northeast)	Region VI (Southeast)	Region VII (Black Sea)	Region VIII (Central East)	Region IX (Central South)	Total for Turkey
1961	1,949	532	511	626	369	823	377	753	1,777	7,717
1962	1,932	539	537	678	354	893	373	727	1,767	7,800
1963	1,922	534	525	661	345	862	397	800	1,804	7,850
1964	1,925	543	544	685	353	830	379	781	1,830	7,870
1965	1,889	566	547	684	350	894	363	816	1,791	7,900
1966	1,904	555	450	716	370	901	403	787	1,865	7,951
1967	1,926	539	566	714	340	924	384	778	1,828	7,999
1968	1,963	578	581	741	389	948	401	798	1,851	8,250
1969	2,066	613	588	815	401	988	445	824	1,920	8,660
1970	2,062	570	535	909	426	1,021	427	785	1,564	8,299
1971	2,088	633	555	818	436	1,087	432	799	1,858	8,706

Source: State Institute of Statistics, Turkey, Agricultural Structure and Production, 1961-72.

TABLE A.9

Hectares in Wheat for Main Areas of Turkey
(thousand hectares)

Year	Central Plateau Area	Thrace Area	Coastal Area, Including South Marmara, Aegean, and Mediterranean Regions Minus Adana Province	Adana Province
1961	2,244	334	1,184	150
1962	2,199	362	1,233	157
1963	2,216	351	1,232	135
1964	2,197	362	1,258	151
1965	2,174	370	1,275	150
1966	2,183	264	1,294	161
1967	2,178	374	1,284	160
1968	2,199	393	1,332	173
1969	2,320	391	1,404	219
1970	2,282	345	1,373	293
1971	2,315	373	1,406	225

Source: State Institute of Statistics, Turkey, Agricultural Structure and Production, 1961-72.

TABLE A.10

Wheat Production in the Nine Agricultural Regions of Turkey
(thousand tons)

Year	Region I (Central North)	Region II (Aegean)	Region III (Marmara)	Region IV (Mediterranean)	Region V (Northeast)	Region VI (Southeast)	Region VII (Black Sea)	Region VIII (Central East)	Region IX (Central South)	Total for Turkey
1961	1,740	606	555	818	171	778	368	630	1,335	7,001
1962	1,998	682	653	960	394	1,074	373	719	1,598	8,451
1963	2,553	657	530	832	363	1,085	456	947	2,576	9,999
1964	2,080	645	684	880	354	521	452	977	1,706	8,299
1965	2,038	685	759	821	331	748	405	801	1,913	8,501
1966	2,470	705	598	902	372	994	452	927	2,179	9,599
1967	2,450	737	811	988	388	871	486	811	2,458	10,000
1968	2,266	832	669	1,038	288	859	429	785	2,354	9,520
1969	2,428	1,002	785	1,409	295	813	510	793	2,465	10,500
1970	2,410	982	864	1,583	320	788	569	732	1,749	9,997
1971	3,266	1,174	1,096	1,831	408	1,081	564	1,008	3,070	13,498

Source: State Institute of Statistics, Turkey, Agricultural Structure and Production, 1961-72.

TABLE A.11

Wheat Production for Main Areas of Turkey
(tons)

Year	Central Plateau Area	Thrace Area	Coastal Area, Including South Marmara, Aegean, and Mediterranean Regions Minus Adana Province	Adana Province
1961	1,648,639	304,183	1,443,036	231,000
1962	1,913,122	412,833	1,641,933	240,000
1963	3,196,765	304,945	1,579,067	135,390
1964	2,113,418	435,000	1,521,034	253,000
1965	2,397,687	497,299	1,530,083	237,158
1966	2,731,752	315,832	1,668,366	221,390
1967	3,096,927	521,089	1,783,454	231,494
1968	2,817,947	352,826	1,869,162	316,425
1969	3,013,010	478,744	2,155,186	562,126
1970	2,543,700	529,901	2,228,419	671,546
1971	3,907,812	762,456	2,617,730	720,093

Source: State Institute of Statistics, Turkey, Agricultural Structure and Production, 1961-72.

TABLE A.12

Relative Importance of Wheat Production for the Nine Agricultural Regions of Turkey

Region	Average Annual Percent of Total Area Sown in Wheat			Average Annual Percent of Total Turkish Wheat Area			Average Annual Percent of Total Turkish Wheat Production		
	1961-65	1966-68	1969-71	1961-65	1966-68	1969-71	1961-65	1966-68	1969-71
I (North Central)	64.3	66.6	69.1	24.6	23.9	23.7	24.5	24.7	24.4
II (Aegean)	31.2	31.6	33.5	6.9	6.9	6.9	7.7	7.8	9.5
III (Marmara)	40.0	40.9	42.1	6.8	6.6	6.4	7.5	7.1	8.3
IV (Mediterranean)	40.1	41.9	49.0	8.5	9.0	9.7	10.1	10.1	14.5
V (Northeast)	55.4	54.1	55.8	4.5	4.5	4.8	3.8	3.6	3.1
VI (Southeast)	65.0	68.0	69.6	11.0	11.5	11.9	9.9	9.4	8.1
VII (Black Sea)	31.5	32.7	34.9	4.8	4.9	5.0	8.1	4.7	5.0
VIII (Central East)	61.1	62.7	64.3	9.9	9.8	9.2	9.6	8.7	7.6
IX (Central South)	58.1	58.0	56.1	22.9	22.9	20.3	21.5	24.0	22.0

Source: Calculated by the author.

TABLE A.13

Relative Importance of Wheat Production for Main Areas of Turkey

Area	Average Annual Percent of Total Area Sown in Wheat			Average Annual Percent of Total Turkish Wheat Area			Average Annual Percent of Total Turkish Wheat Production		
	1961-65	1966-68	1969-71	1961-65	1966-68	1969-71	1961-65	1966-68	1969-71
Central Plateau	66.5	67.9	67.7	28.2	27.1	26.3	26.5	29.7	28.5
Thrace	41.8	41.8	47.7	4.6	4.3	4.2	4.6	4.1	5.3
Coastal area, including south Marmara, Aegean, and Mediterranean regions, minus Adana Province	37.1	38.5	39.6	15.8	16.2	15.9	18.1	18.3	21.1
Adana Province	27.1	28.3	43.4	1.9	2.0	2.8	2.6	2.6	5.9

Source: Calculated by the author.

TABLE A.14

Wheat Yields in the Nine Agricultural Regions of Turkey
(kilograms per hectare)

Year	Region I (Central North)	Region II (Aegean)	Region III (Marmara)	Region IV (Mediterranean)	Region V (Northeast)	Region VI (Southeast)	Region VII (Black Sea)	Region VIII (Central East)	Region IX (Central South)	Total for Turkey
1961	892	1,135	1,085	1,306	462	946	974	836	751	907
1962	1,033	1,266	1,216	1,414	1,114	1,202	999	987	904	1,083
1963	1,328	1,230	1,009	1,258	1,052	1,258	1,150	1,184	1,427	1,274
1964	1,081	1,188	1,256	1,285	1,003	628	1,195	1,251	932	1,055
1965	1,079	1,212	1,386	1,200	946	836	1,115	982	1,068	1,076
1966	1,298	1,272	1,329	1,260	1,007	1,104	1,122	1,177	1,168	1,207
1967	1,272	1,369	1,433	1,382	1,142	942	1,266	1,042	1,344	1,250
1968	1,154	1,441	1,151	1,401	342	905	1,070	983	1,272	1,154
1969	1,176	1,635	1,335	1,729	781	848	1,148	1,003	1,308	1,212
1970	1,167	1,725	1,616	1,742	752	773	1,332	933	939	1,205
1971	1,564	1,854	1,973	2,239	937	994	1,333	1,261	1,652	1,550

Source: State Institute of Statistics, Turkey, Agricultural Structure and Production, 1961-72.

TABLE A.15

Wheat Yields for Main Areas of Turkey
(kilograms per hectare)

Year	Central Plateau Area	Thrace Area	Coastal Area, Including South Marmara, Aegean, and Mediterranean Regions Minus Adana Province	Adana Province
1961	734	910	1,218	1,535
1962	870	1,139	1,331	1,521
1963	1,442	867	1,281	998
1964	962	1,199	1,209	1,673
1965	1,103	1,343	1,199	1,576
1966	1,251	1,193	1,289	1,373
1967	1,422	1,391	1,389	1,445
1968	1,281	896	1,403	1,829
1969	1,299	1,222	1,535	2,558
1970	1,114	1,532	1,623	2,285
1971	1,688	2,042	1,861	3,191

Source: State Institute of Statistics, Turkey, Agricultural Structure and Production, 1961-72.

TABLE A.16

Indexes of Wheat Yields for the Nine Agricultural Regions of Turkey

Year	Region I (Central North)	Region II (Aegean)	Region III (Marmara)	Region IV (Mediterranean)	Region V (Northeast)	Region VI (Southeast)	Region VII (Black Sea)	Region VIII (Central East)	Region IX (Central South)
1961	100	100	100	100	100	100	100	100	100
1962	116	112	112	108	241	127	103	118	120
1963	149	108	93	96	228	133	118	142	190
1964	121	105	116	98	217	66	123	150	124
1965	121	107	128	92	205	88	114	117	142
1966	146	112	122	96	218	117	115	141	156
1967	143	121	132	106	247	100	130	125	179
1968	129	127	106	107	74	96	110	118	169
1969	132	144	123	132	169	90	118	120	174
1970	131	152	149	133	163	82	137	112	125
1971	175	163	182	171	203	105	137	151	220

Source: Calculated by the author.

TABLE A.17

Indexes of Wheat Yields for Main Areas of Turkey

Year	Central Plateau Area	Thrace Area	Coastal Area, Including South Marmara, Aegean, and Mediterranean Regions Minus Adana Province	Adana Province
1961	100	100	100	100
1962	119	125	109	99
1963	196	95	105	65
1964	131	132	99	109
1965	150	148	98	103
1966	170	131	106	89
1967	194	153	114	94
1968	175	98	115	119
1969	177	134	126	167
1970	152	168	133	149
1971	230	224	153	208

Source: Calculated by the author.

TABLE A.18

Indexes of Wheat Production for the Nine Agricultural Regions of Turkey

Year	Region I (Central North)	Region II (Aegean)	Region III (Marmara)	Region IV (Mediterranean)	Region V (Northeast)	Region VI (Southeast)	Region VII (Black Sea)	Region VIII (Central East)	Region IX (Central South)
1961	100	100	100	100	100	100	100	100	100
1962	115	113	118	117	230	138	101	114	120
1963	147	108	95	102	212	139	124	150	193
1964	120	106	123	108	207	67	123	155	128
1965	117	113	137	100	194	96	110	127	143
1966	142	116	108	110	218	128	123	147	163
1967	141	122	146	121	227	112	132	129	184
1968	130	137	121	127	168	110	117	125	176
1969	140	165	141	172	173	104	139	126	185
1970	138	162	156	194	187	101	155	116	131
1971	188	194	197	224	239	139	153	160	230

Source: Calculated by the author.

TABLE A.19

Indexes of Wheat Production for Main Areas of Turkey

Year	Central Plateau Area	Thrace Area	Coastal Area, Including South Marmara, Aegean, and Mediterranean Regions Minus Adana Province	Adana Province
1961	100	100	100	100
1962	116	136	114	104
1963	194	100	109	59
1964	128	143	105	110
1965	145	163	106	103
1966	166	104	116	96
1967	188	171	124	100
1968	171	116	130	137
1969	183	157	149	243
1970	154	174	154	291
1971	237	251	181	312

Source: Calculated by the author.

TABLE A.20

Prices of Wheat and Competing Crops in Turkey
(kurus per kilogram)

Year	Wheat	Barley	Cotton Lint	Dry Beans	Sunflower	Dried Alfalfa
1961	72.7	53.6	516.5	161.4	110.3	32.2
1962	82.4	59.0	524.2	174.0	119.0	35.4
1963	82.1	58.8	535.0	229.5	131.5	32.8
1964	81.4	58.6	523.2	233.5	134.8	32.2
1965	85.8	65.1	516.6	216.2	172.2	34.8
1966	89.9	71.3	511.7	208.1	167.9	36.7
1967	89.9	70.7	543.4	208.1	195.6	38.4
1968	91.7	74.2	551.7	235.3	176.3	40.7
1969	97.2	80.4	564.6	269.2	177.5	48.9
1970	101.1	83.9	618.8	300.0	200.6	53.9
1971	105.2	86.8	773.0	357.2	215.2	53.5
1972	109.8	93.7	935.2	433.1	267.6	63.0

Note: 100 kurus equal 1 TL.

Sources: AID, Ankara, Economic and Social Indicators--Turkey; and State Institute of Statistics, Turkey, Agricultural Structure and Production, Statistical Yearbook of Turkey, 1973.

TABLE A.21

General Turkish Price Indexes and Price Indexes for Wheat and Competing Crops in Turkey
(1963 = 100.0)

Year	General Wholesale Price Index Turkey	Cost of Living, Istanbul	Cost of Living, Ankara	Wheat	Barley	Cotton Lint	Dry Beans	Sunflower	Dried Alfalfa
1961	90.9*	92.0*	93.8	88.6	91.2	96.5	70.3	83.9	98.2
1962	95.8*	96.0*	96.7	100.0	100.3	98.0	75.8	90.5	107.9
1963	100.0	100.0	100.0	100.0	100.0	100.0	100.0	100.0	100.0
1964	101.2	100.8	101.9	99.1	99.7	97.8	101.7	102.5	98.2
1965	109.4	104.8	107.6	104.5	110.7	96.6	94.2	131.0	106.1
1966	114.7	113.6	113.5	109.5	121.3	95.6	90.7	127.7	111.9
1967	123.4	129.6	120.6	109.5	120.2	101.6	90.7	148.7	117.1
1968	127.3	137.6	125.6	111.7	126.2	103.1	102.5	134.1	124.1
1969	136.5	144.2	132.7	118.4	136.7	113.8	117.3	135.0	149.1
1970	145.7	155.6	148.3	123.1	144.2	134.8	130.7	152.5	164.3
1971	168.9	185.2	180.7	128.1	147.6	144.5	155.6	163.7	163.1
1972	199.3	213.7	208.3	139.8	161.0	175.0	188.7	203.5	192.1

Note: Commodity prices for 1963 (in kurus per kilogram) are wheat, 82.1; barley, 58.8; cotton lint, 535.0; dry beans, 229.5; sunflower, 131.5; and dried alfalfa, 32.8.

*Estimates.

Source: (1) The indexes for the commodity prices are calculated by the author on the basis of the prices listed in Table A.20.
(2) The indexes, AID, Ankara, Economic and Social Indicators--Turkey, and State Institute of Statistics, Turkey, Statistical Yearbook of Turkey, 1973.

TABLE A.22

Wholesale Prices and Retail Prices of Selected Livestock Products in Istanbul
(kurus per kilogram, 100 kurus = 1 TL)

	Wholesale Prices					Retail Prices					
Year	Mutton	Beef	Veal	White Cheese	Eggs	Mutton	Beef boneless	Poultry (dressed)	Eggs (each)	Milk	White Cheese
1961	836.8	477.1	473.2	516.7	303.5	748	657	1,224	27	148	669
1962	653.1	493.2	478.7	517.6	333.4	802	722	1,197	30	159	710
1963	757.1	561.5	573.1	541.2	365.2	863	788	1,257	31	160	732
1964	782.7	623.8	651.7	545.1	377.8	919	842	1,417	32	163	727
1965	813.9	680.2	697.8	617.7	421.6	911	865	1,490	36	158	803
1966	931.6	706.6	738.9	619.9	460.2	1,066	970	1,804	41	176	809
1967	1,055.9	854.5	846.4	742.7	489.4	1,350	1,211	2,021	44	196	1,001
1968	1,128.8	877.9	866.8	849.0	526.1	1,469	1,267	2,250	47	220	1,144
1969	1,125.0	878.5	937.5	893.2	517.8	1,375	1,183	2,188	47	249	1,089
1970	1,253.7	1,030.6	1,078.7	906.7	626.9	1,406	1,293	2,022	53	257	1,114
1971	1,491.6	1,334.7	1,396.2	1,105.1	745.9	1,786	1,769	1,905	64	287	1,356
1972	1,822.0	1,729.7	1,726.7	1,291.9	803.2	2,121	1,960	1,977	67	315	1,564

Source: State Institute of Statistics, Turkey, Monthly Bulletin of Statistics, 1970-IX and 1973-VIII.

TABLE A.23

Index Numbers of Wholesale and Consumer Prices, Turkey
(1963 = 100)

	1965	1966	1967	1968	1969	1970	1971	1972
Wholesale Prices, Turkey								
General	109.4	114.7	123.4	127.3	136.5	145.7	168.9	199.3
Total foodstuffs and Livestock feed	110.3	116.3	124.0	126.4	136.6	141.0	161.0	185.4
Vegetable foodstuffs	104.5	108.4	110.6	113.7	123.9	127.0	138.4	153.4
Animals	116.7	126.5	152.8	148.3	155.3	179.1	255.1	331.5
Animal products	111.9	121.7	138.1	133.6	141.6	150.0	171.8	189.8
Industrial raw material and semifinished products	107.8	112.2	122.3	128.8	136.5	153.6	182.1	222.4
Consumer Prices, Ankara								
General	107.6	113.5	120.6	125.6	132.7	148.3	180.7	208.3
Foodstuffs	110.4	115.1	123.7	129.7	135.6	149.9	175.4	197.3
Cereals and cereal products	117.1	117.8	117.4	121.2	126.7	145.2	168.4	180.0
Meat	129.8	144.7	170.6	180.8	187.6	210.3	256.3	312.6
Milk and dairy products	110.1	113.9	124.5	134.4	143.1	152.3	182.2	200.0
Dried vegetables and fruits	96.6	103.1	99.7	106.2	122.5	125.8	143.4	174.4

Source: State Institute of Statistics, Turkey, Statistical Yearbook of Turkey, 1973.

BIBLIOGRAPHY

Agricultural Bank of Turkey. Agricultural Credit and Sales Cooperatives. Published annually, Ankara.

Aksoz, Ibrahim. "Under-employment in Agriculture in Turkey and Fuller Utilization of Rural Manpower." Paper presented at August 1964 conference on "Agricultural Aspects of Economic Development," ESSCB, Istanbul.

Aktan, Resat. "Agricultural Problems of Turkey." Mediterranea 12 (October-December 1966): 266-75.

______. Analysis and Assessment of the Economic Effects of Public Law 480 Title I Program: Turkey. Ankara: Ankara University Press, 1966.

______. "Basic Characteristics of Turkish Agriculture and Problems of Productivity." Paper presented at the October 1972 conference on "Problems of Increasing Productivity in Turkish Agriculture and the International Experience," ESSCB, Istanbul.

______. "Economic Effects of U.S. Agricultural Commodity Aid Program in Turkey." Turkish Yearbook of International Relations. Ankara, 1964.

______. "The Nature and Importance of Technical Change in Agricultural Production." Paper presented at June 1967 conference on "Increasing the Value of Agricultural Products," ESSCB, Istanbul.

______. "Problems of Land Reform in Turkey." The Middle East Journal 20, no. 3 (Summer 1966): 317-24.

Aral, Namik Zeki. "Agricultural Cooperatives in Turkey," Section 2 of Seminar Report on Economic Development and Cooperatives, Mediterranean Social Sciences Research Council and Institute of Economica and Sociology, Istanbul, February 1965. Istanbul University Publication No. 1119, 1965.

______. "Agricultural Credits and Credit Establishments in Turkey." Paper presented at August 1964 conference on "Agricultural Aspects of Economic Development," ESSCB, Istanbul.

Balaban, Ali. "Agricultural Productivity and Irrigation in Turkey." Paper presented at August 1963 conference on "Agricultural Aspects of Economic Development," ESSCB, Istanbul.

Beeley, Brain W. and Proud, R. B. "Technically Trained Manpower in Turkish Agriculture, 1963-1971." In Mediterranean Regional Project: Turkey. OECD, Paris, 1965, pp. 169-89.

Borlaug, Norman E. "Wheat Breeding and Its Impact on World Food Supply." Proceedings of the 3rd International Wheat Genetics Symposium, Canberra, 1968. Australian Academy of Science, pp. 1-36.

______ and Aresvik, Oddvar H. "The Green Revolution--An Approach to Agricultural Development and Some of Its Economic Implications." International Journal of Agrarian Affairs 5, no. 6 (March 1973).

______, Narvaez, M. I., Aresvik, Oddvar, and Anderson, Glenn. "A Green Revolution Yields a Golden Harvest." Columbia Journal of World Business, September/October 1969.

Blolock, James I. "Capital and Finance in Turkish Agriculture." Economic Planning Division, AID, Ankara, May 1969.

Brandow, G. E. "Agricultural Development in Turkey." United States Foreign Operations Administration, Ankara, December 1953.

Bulatayit, Tunker et al. "Income Distribution in Turkey, 1968." Faculty of Political Sciences, Ankara University, Ankara, 1971 (in Turkish).

Bush, Charles D. "Agricultural Mechanization in Turkey: A Preliminary Report." U.S. Operations Mission--Turkey, Food and Agricultural Division, Ankara, December 1960.

Boykurt, Kaya. "Irrigation and Soil Improvement Problems in Agriculture." Paper presented at the October 1972 conference on "Problems of Increasing Productivity in Turkish Agriculture and the International Experience," ESSCB, Istanbul.

Casey, Doyle et al. "A Reorganization of Turkish Agriculture." Project Management Board of the Organization Project Group, Ministry of Agriculture, Ankara, 1964.

Cohn, Edwin J. Turkish Economic, Social and Political Change: The Development of a More Prosperous and Open Society. New York: Praeger Publishers, 1970.

_____. "Turkish Forest Products: Problems and Opportunities." AID, Ankara, May 1967.

Cummings, Ralph W., Rodenluser, H. A., and Gibler, J. W. "The Cooperative Wheat Research Program in Turkey: A Report to the Ministry of Agriculture and to The Rockefeller Foundation." Ankara, August 1968.

Demirlicakmak, Ahmet. "The Wheat Project in Turkey." Paper presented at the CENTO Workshop for Agricultural Planners, December 1972, Ankara.

_____. "Turkey Country Report." Paper presented at Third FAO Wheat Seminar, April-May 1970, Ankara.

Doren, Van P. J. "Organization and Operation of Agricultural Credit Marketing and Village Development Cooperatives." FAO, No. TA 2557, Rome, 1968.

Economic and Social Studies Conference Board. "Agricultural Aspects of Economic Development." Report of the Third International Conference, Istanbul, August 1964.

_____. "Foreign Trade and Economic Development." Report of the International Seminar on Foreign Trade and Economic Development, Istanbul, August-September 1968.

Elkinton, Charles M. "Prospects for Turkish Agriculture." A Report by AID Study Team to the Minister of Agriculture. Ankara, December 1966.

Food and Agricultural Organization. Country Report for Turkey. FAO Mediterranean Development Project, Rome, 1959. (Report within the FAO Mediterranean Development Project covering agriculture and forestry in relation to economic growth.)

_____. "Report to the Government of Turkey: Organization of Agricultural Credit, Marketing and Village Development Cooperatives." No. TA 2557, Rome, 1968.

Frey, Fredrick W. "Regional Variations in Rural Turkey." Cambridge, Mass.: MIT Center for International Studies, Rural Development Research Project Report No. 4, December 1966.

Forker, Olan D. "Agricultural Prices Intervention by the Government of Turkey." AID, Turkey, August 1967. (mimeographed).

_____. "Agricultural Price Policy in Turkey. A Collection of Works." Volumes 1 and 2. Economic Staff Papers, Discussion Paper No. 3, AID, Ankara.

_____. "Marketing, Subsidies and Price Policy." Paper presented at the October 1972 conference on "Problems of Increasing Productivity in Turkish Agriculture and the International Experience," ESSCB, Istanbul.

Gunes, Turan. "Marketing of Beef and Cow Milk in Turkey." In Proceedings of the International Symposium on Cattle Rearing, Centre National des Exposition et Concours Agricoles, Paris, 1967.

_____. "Price Policy of Agricultural Products in Turkey." Paper presented at Euro-Mediterranean Symposium on Research in Agricultural Economics and Rural Sociology, Centro de Estudos Economica Agraria, Lisbon, October 1966.

Hafiz, Abdul. "Report on Cereal Improvement and Production in Turkey," FAO, Cereal Improvement and Production Project, Cairo, July 1971.

_____. "Report on Wheat and Barley Improvement in Turkey." FAO, Near East Wheat and Barley Improvement and Production Project, Cairo, August 1966.

Harness, Vernon L. "Cotton in Turkey." USDA, Foreign Agricultural Service (FAS-M-179), 1966.

Helburn, Nicholas. "A Stereotype of Agriculture in Semi-arid Turkey." Geographical Review 45, no. 3 (1955): 375-84.

_____. "Some Trends in Turkish Agriculture." Dogus Malbassi, Ankara, 1953.

Henning, George F. "Livestock and Meat Marketing in Turkey." AID, Ankara, 1960.

Hershlag, Zvi Y. Turkey: The Challenge of Growth. Leiden: E. J. Brill, 1968.

Hill, John M. "A Discussion Paper on the Need for Re-organization of Turkey's Fertilizer Industry." AID, Ankara, February 1971.

Hirsch, Eva. "Poverty and Plenty on the Turkish Farm." New York: 1970.

Humphrey, L. M. "Mexican Wheat Comes to Turkey." Food and Agriculture Division, AID, Turkey, April 1969.

Ilkin, S., and Inanc, E., eds. "Planning in Turkey." Faculty of Administrative Science, Middle East Technical University, Ankara, 1967.

International Bank for Reconstruction and Development. "The Economy of Turkey: An Analysis and Recommendation for a Development Program." Washington, D.C.: 1951.

International Maize and Wheat Improvement Center. "Annual Reports." Mexico, D.F., Mexico.

Jackson, Thomas L., Kronstad, Warren E., and Vogel, Orville A. "Expansion of Wheat Production in Turkey, II." AID, Turkey, July 1968.

Jackson, Thomas L. and Kronstad, Warren E. "Final Report on Expansion of Wheat Production in Turkey." AID, Turkey, August 1967.

Johnson, Harvey P. H. "High Yielding Winter Wheat for Turkey: A Progress Report." AID, Turkey, August 1971.

Keles, Rusen Y. "Regional Disparities in Turkey." Reprint by the Faculty of Political Science, Ankara University, Ankara, 1964.

_____. "Urbanization and Balanced Regional Development in Turkey." Ekistics 22, no. 129 (August 1966).

Kiray, M. and Hinderink, Jon. "Interdependencies Between Agronomic Development and Social Change: A Comparative Study Conducted in the Cukurova Region of Southern Turkey." Journal of Development Studies (July 1968).

Kolars, John F. "Tradition, Season, and Change in a Turkish Village." Department of Geography Research Paper, no. 82, University of Chicago Press, Chicago, 1963.

Klatt, Art. "Report on Trip to Turkey, May-June 1971." International Maize and Wheat Improvement Center, Mexico, D. F. Mexico, 1971.

Lewis, Bernard. "The Emergence of Modern Turkey." London: Royal Institute of International Affairs, 1961.

Mann, Charles K. "Irrigation and Soil Improvement Problems in Agriculture." Paper presented at the October 1972 conference on "Problems of Increasing Productivity and the International Experience," ESSCB, Istanbul.

Money-Kryle, A. F. "Agricultural Development and Research in Turkey." American University of Beirut, Faculty of Agricultural Sciences, Publication no. 4, edited by R. H. Porter, Beirut, December 1967.

Oakes, Harvey. "The Soils of Turkey." USFOA and Ministry of Agriculture, Ankara, 1957.

Organization for Economic Cooperation and Development. "Agricultural Projections for 1975 and 1985" with mimeographed country studies for each OECD country. OECD, Paris, 1969.

_____. "Regional Planning in Turkey: The Marmara Pilot Project." OECD, Paris, 1962.

_____. "Turkey: Aims Pursued by the Government in Directing Cereal Production and Methods Used." Third Report on the Agricultural Policies in Europe and North Africa, OEEC, Paris, 1958.

_____. "Turkey, in Low Incomes in Agriculture: Problems and Policies." OECD, Paris, 1964.

_____. "Turkey." In Regional Economic Planning: Techniques and Analysis. Edited by W. Isard and J. H. Cumberland. EPA/OEEC, Paris, 1961.

_____. "Turkey, The Marketing of Fruits and Vegetables in Syria and Turkey." EPA/OEEC, Paris, 1960.

Okyar, Osman. "Agricultural Price Policy: Turkish Experience." Paper presented at August 1964 conference on "Agricultural Aspects of Economic Development," ESSCB, Istanbul.

_____. "Development of Eastern Turkey." In Regional Planning, Local Government, and Community Development in Turkey. TSHP, Publication No. 3, Ankara, 1966, pp. 15-27.

Palmer, Z. Edgar. "Agriculture in Turkey--Long Term Projections of Supply and Demand." Roberts College, School of Business Administration and Economics, Istanbul, 1966.

Racham, T. S. "Development of Farm Management Activity in the Ministry of Agriculture, Ankara." FAO EPA, Report No. 1847, Ankara, 1964.

Sari, E. E. "A Report on Wheat Pathology to the Minister of Agriculture." Ankara, August 1970.

_____. "Questions and Answers to Points Pertaining to the Turkey Wheat Program and the Middle East Area." New Delhi, September 1970 (mimeographed).

Samuels, J. Kenneth. "Improving the Marketing of Fresh Fruits and Vegetables in Turkey." AID, Ankara, October 1960.

Sireli, Aykut. "Efficiency in the Domestic Marketing of Turkish Agricultural Products." Paper presented at June 1967 conference on "Increasing the Value of Agricultural Products," ESSCB, Istanbul.

Sonmez, Necmi. "Existing Land Ownership and Land Use Conditions in Turkey." Yearbook of the Faculty of Agriculture, University of Ankara, 1963.

State Institute of Statistics, Turkey. Agricultural Structure and Production. Published annually, Ankara.

_____. Statistical Yearbook of Turkey. Published annually, Ankara.

_____. Summary of Agricultural Statistics. Published annually, Ankara.

State Planning Organization, Turkey. "First Five-Year Development Plan, 1963-67." Ankara, 1962 (English translation, 1963).

_____. "Second Five-Year Development Plan, 1968-72." Ankara, 1967 (English translation, 1969).

Stirling, Paul. Turkish Village. London: Weidenfeld and Nicolson, 1965.

Taylor, Donald C. "Research on Agricultural Development in Selected Middle Eastern Countries." Agricultural Development Council, New York, 1968.

Tekeli, I. "Regional Planning in Turkey and Regional Policy in the First Five-Year Development Plan." In Planning in Turkey: Selected Papers. Edited by S. Ilkin and E. Inanc. METU, Faculty of Administrative Sciences, Publication No. 9, Ankara, 1967.

ABOUT THE AUTHOR

ODDVAR ARESVIK is currently with the International Bank for Reconstruction and Development, serving as Agricultural Economics Advisor to the government of Jordan. A distinguished agricultural economics specialist, Mr. Aresvik has been with the Ford Foundation as an agricultural economist in a Middle East development program and as an advisor to the Pakastani government.

From 1950 to 1963, Mr. Aresvik was Professor and Chairman of the Department of Agricultural Economics and Rural Sociology and the Agricultural University of Norway. During his tenure at the University, Mr. Aresvik also served as an advisor to the government of Norway in the field of agricultural planning and development.

Mr. Aresvik received his Cand.Agric. degree from the Agricultural University of Norway, and his Cand.Econ. degree from Oslo University.

RELATED TITLES
Published by
Praeger Special Studies

AGRICULTURAL DEVELOPMENT PLANNING:
Economic Concepts, Administrative Procedures
and Political Process
Willard W. Cochrane

AGRICULTURAL DEVELOPMENT OF IRAN
Oddvar Aresvik

AGRICULTURAL MODERNIZATION THROUGH
PRODUCTION CONTRACTING: The Role of the
Fruit and Vegetable Processor in Mexico
and Central America
J. David Morrissy

DEVELOPMENT IN RICH AND POOR COUNTRIES:
A General Theory with Statistical Analyses
Thorkil Kristensen

75 02352